What You'll Find on the CD-ROM

This full-featured interactive CD-ROM puts you in charge of a fantastic voyage through the innermost layers of the human body. There, you'll view over an hour of video and 200 illustrations and hear stereo sound that'll guarantee you'll find out exactly what you're made of.

This exciting 3D laboratory environment is easy to use, letting you explore the complex systems and organs of the body with a simple click of the mouse. If you can click, you can:

❖ Fly through animated multidimensional tours of your body's systems, including circulation, respiration, thought and action, locomotion, digestion, reproduction, and many others.

❖ View numerous video clips of medical experts discussing health disorders and wellness issues.

❖ Access health advice from a medical glossary, online references, up-to-date medical articles, anatomy charts, and even nursing expert and radio personality Pat Carroll, R.N.

❖ Test your skills in an entertaining scavenger hunt in and around the laboratory, where your findings will reward you with fun multimedia surprises, such as a navigable heart.

Focused on the body's processes—not just its parts—this highly interactive CD-ROM presents a compelling, progressive multimedia anatomy adventure for the whole family. Combine it with the content-rich, beautifully illustrated book or view each alone. Either way, you'll surely be amazed at *How Your Body Works*.

PC System Requirements

❖ 486 SX 25 MHz (33 MHZ recommended)
❖ 8 MB RAM or greater
❖ 4 MB available on hard drive (10 MB recommended)
❖ Microsoft Windows 3.1 or higher
❖ Double-speed CD-ROM drive
❖ Super VGA (640 x 480 resolution)*
❖ 256 (8-bit) color (16-bit recommended)
❖ Sound Blaster sound card, or compatible*
❖ Mouse

Current sound card drivers and video drivers recommended

This CD-ROM is also available in Macintosh format. If you would like to receive the Macintosh version, please return the PC version, with your name and address to:

Disk Exchange
Ziff-Davis Press
5903 Christie Ave.
Emeryville, CA 94608

Macintosh System Requirements

❖ 68030 25 Mhz or faster
❖ 8 MB RAM (4.2 MB available)
❖ 4 MB available on hard drive (10 MB recommended)
❖ System 7.1 (or higher)
❖ Double-speed CD-ROM drive
❖ 13-inch color monitor
❖ 256 (8-bit) color (16-bit recommended)
❖ Mouse

COMPLETE INSTRUCTIONS FOR INSTALLING AND RUNNING THE CD-ROM ARE AT THE BACK OF THIS BOOK.

HOW YOUR BODY WORKS

INCLUDES INTERACTIVE CD-ROM

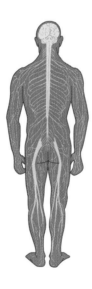

HOW YOUR BODY WORKS DESCRIBES THE HUMAN BODY IN GENERAL AND DISCUSSES MANY COMMON CONDITIONS AND DISORDERS. IT TRIES TO MAKE YOU A SMARTER CONSUMER OF HEALTH SERVICES AND PRODUCTS, BUT IT DOES NOT OFFER MEDICAL ADVICE AND IS NOT A SUBSTITUTE FOR MEDICAL CARE OR SUPERVISION. CONSULT A PHYSICIAN ABOUT ALL YOUR SPECIFIC HEALTH CONCERNS.

HOW YOUR BODY WORKS

INCLUDES INTERACTIVE CD-ROM

SUSAN L. ENGEL-ARIELI, M.D.

Development Editors	Mary Johnson and Cheryl Holzaepfel
Copy Editor	Mary Johnson
Technical Reviewer	Bryce Palchick, M.D.
Proofreader	Carol Burbo
Cover Design	Regan Honda
Book Design	Carrie English
Digital Illustration	Dave Feasey and Sarah Ishida
Word Processing	Howard Blechman
Page Layout	Bruce Lundquist
Indexer	Valerie Robbins

Ziff-Davis Press books are produced on a Macintosh computer system with the following applications: FrameMaker®, Microsoft® Word, QuarkXPress®, Adobe Illustrator®, Adobe Photoshop®, Adobe Streamline™, MacLink®*Plus*, Aldus® FreeHand™, Collage Plus™.

If you have comments or questions or would like to receive a free catalog, call or write:
Ziff-Davis Press
5903 Christie Avenue
Emeryville, CA 94608
1-800-688-0448

ISBN 1-56276-286-9

Manufactured in the United States of America
♲ This book is printed on paper that contains 50% total recycled fiber of which 20% is de-inked postconsumer fiber.
10 9 8 7 6 5 4 3 2 1

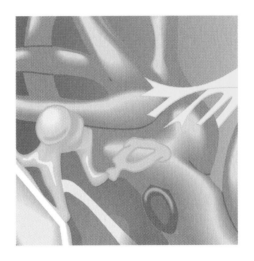

This book is dedicated to my mother, Marion Carpenter Kasper, for her herculean efforts in helping me attain many of my goals in medicine, and for her unconditional love and support.

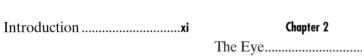

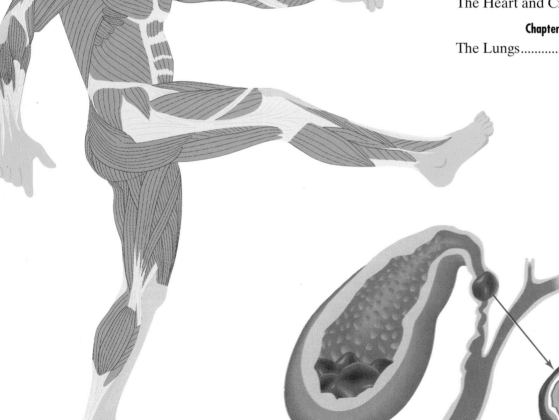

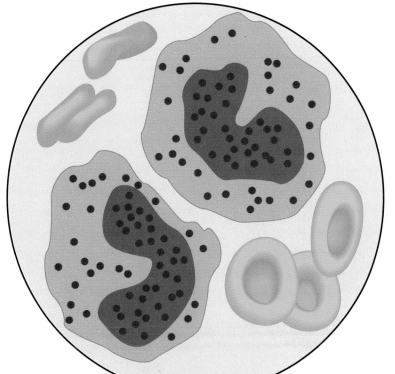

No woman or man stands alone in the production of an undertaking of this magnitude. This book was molded by many hands and minds. Therefore, I would like to express my deepest gratitude to the many people who have journeyed down the path with me, and helped in the production of this book.

First, I would like to thank Cindy Hudson, President of Ziff-Davis Press, for her encouragement and motivation in this book. Eric Stone, Acquisitions Editor, was also extremely helpful, understanding, and was always available to answer questions and concerns.

My heartfelt thanks go out to my editors (in the order of coming onto the project): Melinda Levine for her exacting attention to details and asking questions, and Sharon Maddox for her editorial style and flair. Cheryl Holzaepfel, Editor-in-Chief, was involved every step of the way, and I am grateful for, and thank her for her help with the illustration concepts and more. Mary Johnson, who participated in the majority of the editorial assistance for the book, is also thanked for providing editorial guidance and questions and ideas regarding materials to include in the book. Last, but not least, my gratitude goes to Ami Knox, Project Coordinator, for her perseverance, skills, and patience in the revision process.

Dr. Bryce Palchick, of Pittsburgh, Pennsylvania, a reviewer for this book, was an invaluable source of ideas, information, and corrections, particularly since my deadlines entailed that the writing had to be done at race car speed. This book was enriched by his input, and he is appreciated and greatly thanked.

Immense thanks also go to the computer art illustrators at Ziff-Davis Press, Dave Feasey and Sarah Ishida, for their computer graphic masterpieces. This was no small feat, and I congratulate and thank all the illustrators very much.

I am also very grateful to many other people at Ziff-Davis Press for their help behind the scenes. They include, in alphabetical order: Howard Blechman (for word processing), Carol Burbo (for proofreading), Bruce Lundquist (for page layout), and Cori Pansarasa (for proofreading and editorial assistance).

Much appreciation goes to Lutheran General Hospital and Dr. John P. Anastos, D.O., radiologist at Lutheran General Hospital for his radiologic interpretations and radiologic films for this book. Additional thanks go to GE Medical Systems and Peg Ireland and Brian Johnson of GE Medical Systems, Milwaukee, Wisconsin for providing additional radiologic films; and to Jim Gardner, VP Marketing of Neuro-Com International of Clackamas, Oregon for his assistance and photographs.

Gratitude also goes to Bill Kasper for his patience and encouragement, and for helping me with related tasks in researching and preparing this book.

Lastly, I would like to thank my husband, Udi Arieli, for his great computer assistance and critiques for this book; for his patience during this undertaking; and for his loving support during the writing of this book.

The human body was created for the sake of the soul and stands erect among other animals....
Man alone has the power of thought; likewise he alone remembers.

—Alessandro Benedetti (1450–1512),
The History of the Human Body

This book will take you on a wonderful journey through your body. It was written for anyone with a desire or need to know more about the body and how it works. If you struggled through high school biology, or if you find the body puzzling and complicated, this book is for you. If you want to learn more about this wondrous machine and how to keep it healthy, this book is for you, too. If you (or someone you know) have a health problem and want to learn more about it, this book can also help you.

People naturally crave knowledge about the body. A look at the mechanics of the human body provides a complex and beautiful panorama—but we are still a long way from completely understanding what we see. Each of us contains a magical combination of cells that forms tissues and organs and that allows us to function in the world around us. Imagine for a moment all that is going on in your body right now! As your eyes take in the images on this page, electrical impulses are being sent to your brain to interpret the information; your hand muscles are preparing to turn the page; your heart is beating; your lungs are expanding; new blood cells are being produced; billions of other cells are being created, living, or being made extinct.

This book contains 16 chapters about the brain, eyes, ears, nose, tongue, heart, lungs, bones, muscles, arms, legs, abdomen, glands, and sex organs and about pregnancy and other topics. The material is presented in simple language and with an abundance of illustrations that clearly show you how your body works. Although far from comprehensive, this book will provide you what you need for a fundamental—and practical—understanding.

In addition to showing the workings of your body, each chapter also discusses some common maladies, with accompanying illustrations. A sampling of the topics includes tension and migraine headaches, nearsightedness, skin cancers, osteoporosis, ankle sprains, stomach ulcers, the effects of smoking and excessive alcohol consumption, diabetes, breast cancer, prostate cancer, and other common problems. Again, the material is not all-encompassing—it's not meant to enable you to diagnose an illness (that should be left up to your own healthcare professional, of course). What it will provide is a sense of the magnitude, symptoms, and workings of a disease or condition. Modes of treatment have not been included as they are beyond the scope of this book; new discoveries are constantly being made, and no two illnesses are exactly the same or use exactly the same treatments.

It is the author's hope that you will find this book useful and interesting, for that is what we have all strived to obtain. We hope to have answered many of your questions—and to have brought forth some new ones, as well.

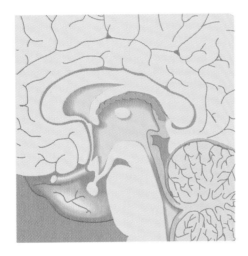

The Brain and Nervous System

THE NERVOUS SYSTEM consists of the brain, spinal cord, peripheral nerves emanating from the spinal cord, and the autonomic nervous system, which controls involuntary functions in the body. The brain and nervous system together provide most of the master controls of the body and regulate thinking, sleeping, perception of pain, breathing, muscle contractions, secretion of some glands, temperature, the arousals and sensations of sex, and more. Furthermore, they process and store information.

One of the major functions of the brain and nervous system is to process incoming information in such a way that appropriate responses occur. Most of the information received by the brain is discarded as insignificant. If it did not do this, you would be bothered by the pressure of clothes on your body and disturbed by all the objects in your field of vision and all the noises around you. Also, only a small amount of important incoming information produces an immediate response. Most information is stored in the cerebral cortex of the brain as memory. Once memories are stored, they become part of a processing chain that channels and integrates them into new thinking information.

The brain has three major components: the cerebrum, cerebellum, and brain stem. The cerebrum receives information, thinks, processes, and sends out information. The cerebellum is essential for good balance and coordination. The brain stem connects the cerebrum to the spinal cord. The brain stem is very complex anatomically, but it primarily consists of the medulla oblongata, which regulates functions such as respiration, heart rate, and blood flow; the pons, which controls eye movements and pupil size; and the midbrain, which controls eye, neck, and head movements.

Sleep is indispensable to our existence and is controlled within the medulla, pons, and other areas of the brain such as the hypothalamus. Our sleep-wake cycle appears to be age related. Each day, a newborn baby sleeps between 16 and 20 hours; a teenager sleeps about 7.5 hours, and an older adult, about 6.5 hours. We know that there are five stages of sleep, based on studies of brain waves using a test called an EEG (electroencephalogram) and on clinical studies. When you first begin to sleep, your muscles relax and eye movements become slow and rolling in nature; this is known as stage 1 sleep. Stages 2, 3, and 4 continue to produce deeper levels of sleep; stage 4 is the deepest. The fifth stage of sleep, called REM (rapid eye movement), is where complex visual dreaming occurs.

Pain is also regulated by the brain. Different tissues in your body have different responses to injury. Pain receptors on the skin respond to most types of skin injury, such as pricking, cutting, and crushing. However, these same stimuli produce no pain in other tissues such as the stomach, intestines, joints, or brain. Instead, pain in the stomach, intestines, or joints is induced by distention, spasm, or inflammation. Pain in the head is produced from distention or inflammation in the accompanying blood vessels or nerves.

The brain controls, and is essential for, the regulation of breathing mechanisms. The respiratory center is located in the medulla oblongata and pons within the brain stem. Inspiratory and expiratory nerves, which control inhalation and exhalation, intermingle in this area. High concentrations of carbon dioxide, and, to a lesser extent, high acid content in the blood stream stimulate your taking a breath to draw more oxygen into the body. Increased levels of carbon dioxide in the blood, detected by the respiratory center of the brain, stimulates the inspiratory nerves and inhibits the expiratory nerves, which sends signals to the chest to increase inspiration and breathe. Additional receptors in the carotid arteries in the neck and in the aorta in the chest also help in regulating breathing. When the carbon dioxide content is sufficiently decreased, the inspiratory nerves are inhibited, then the expiratory nerves are freed from inhibitory signals, which creates expiration, and there is a release of carbon dioxide from the body.

Diseases and malfunctions of the brain have the potential to cause more damage and suffering than those that occur in many other parts of the body. Of all the painful states that afflict us, headaches are the most frequent. The brain tissue itself is almost insensitive to pain, but the blood vessels, nerves, and membranes are pain-sensitive.

It is estimated that nine out of ten people suffer at least one headache per year, and an estimated 40 million Americans suffer from chronic headaches. The most common of these are tension headaches, which generally localize in the forehead and back of the neck, and feel like a tight band or vise wrapped around your head.

The next most common headache is the migraine, affecting some 25 million Americans, of which 75% are women. Typically, the most frequent type of migraine, called common migraine, feels like a hammer pounding inside your head, often behind one eye. As the headache progresses, it can become a dull generalized ache with a sensitive scalp. Nausea and vomiting may occur. A different form of migraine (classic migraine) can begin first with flashes of light, and progress to transient numbness, weakness, vertigo, or speech difficulties. It is believed that migraines may arise when nerves surrounding the blood vessels around the brain discharge pain signals, when dilated blood vessels prompt nerve fibers to generate and release unstable levels of certain chemicals, or when certain chemicals (specifically serotonin) become altered.

One of the most painful of headaches is the cluster headache, which usually affects men, and has been described as feeling like a red hot poker stuck through the eye into the head. It can occur in clusters throughout the day or several days, thus its name. There can be symptom-free intervals of weeks to months until the next cluster descends upon the victim. This type of headache is believed to occur due to a malfunction of the hypothalamus.

Epilepsy is another type of brain disorder, this one being characterized by an out-of-control activity of part or all of the brain and, subsequently, the nervous system. It may be compared to a devastating earthquake, except that this earthquake occurs within the body. Epilepsy creates an uncontrolled discharge of brainwave activity, which can be determined with an EEG. There are two basic types of epileptic seizures: generalized seizures, which originate in the brain stem and other deeper structures (such as the thalamus) and, usually, rapidly spread to all parts of the brain at once, and focal seizures, which involve only a portion of the brain.

Multiple sclerosis is another serious disorder that affects the brain, spinal cord, and/or nerves of the body. Symptoms can include abnormal sensations, weakness of arms or legs, difficulty seeing, bladder control problems, and/or difficulty with balance. Multiple sclerosis is characterized by remissions and recurrences over a period of many years, and because of this, in early stages diagnosis may be difficult. In many cases, the initial symptoms improve somewhat or completely, only to be followed at some later time with the same abnormalities or new ones. Although the cause is not known, it is more prevalent among people living in northern Europe and the northern United States. Also, some hypotheses link it to past viral infections or autoimmune diseases; it can possibly be precipitated by factors such as trauma, pregnancy, or infection.

Strokes, the third most common cause of death in the United States, result from a blockage of the blood supply to a certain part of the brain. More than any other organ, the brain depends on a second-to-second blood supply, and it is because of this that strokes occur so quickly. Strokes are a result of a blood clot in an artery (thrombus) such as a brain artery; a blood clot fragment usually from the heart or a carotid artery of the neck (embolus); or the bursting of a blood vessel (hemorrhage). Symptoms can include paralysis of a part of the body, affected speech, affected eyesight, inability to swallow or talk, and/or disturbances in thinking. Possible risk factors are heart disease, diabetes, high blood pressure, and cigarette smoking.

Senile diseases of the brain currently affect about 4 million people in the United States, and Alzheimer's disease is the most common of these senile disorders. Most Alzheimer's patients are in their 60s or older, but they can also be in their 40s or 50s. The crucial feature of this disease is the disappearance and death of nerve cells in the cerebral cortex, which can produce atrophy of the brain and enlargement of the brain ventricles. Recent studies show that microscopically, brains of Alzheimer's patients have prominent plaques or patches and neurofibrillary tangles (groupings of threadlike filaments in nerve cells), with Beta amyloid protein (an abnormal grouping of a type of protein) deposition, perhaps preceding the development of the disease. Aluminum and a body chemical called apo-protein E may also have a role in development.

It has been stated that in any six-month period, one in five Americans will suffer from some type of mental illness. Depression is the most common, with 9 to 20% of the population experiencing it at any one time. One in four women are affected—compared to one in ten men. The impact on society is great not only in terms of impaired function, but also in lost work time and reduced productivity.

Now, let's take a road trip through the brain, see how it works, and how it is affected by disease.

Inside the Brain and Nervous System

The Brain and Spinal Cord

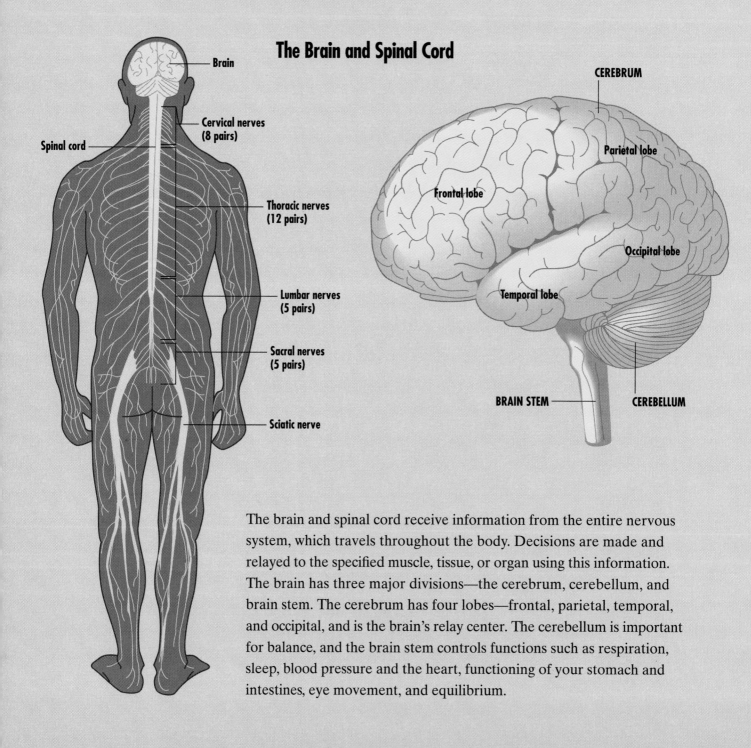

Brain

Cervical nerves
(8 pairs)

Spinal cord

Thoracic nerves
(12 pairs)

Lumbar nerves
(5 pairs)

Sacral nerves
(5 pairs)

Sciatic nerve

CEREBRUM

Parietal lobe

Frontal lobe

Occipital lobe

Temporal lobe

BRAIN STEM

CEREBELLUM

The brain and spinal cord receive information from the entire nervous system, which travels throughout the body. Decisions are made and relayed to the specified muscle, tissue, or organ using this information. The brain has three major divisions—the cerebrum, cerebellum, and brain stem. The cerebrum has four lobes—frontal, parietal, temporal, and occipital, and is the brain's relay center. The cerebellum is important for balance, and the brain stem controls functions such as respiration, sleep, blood pressure and the heart, functioning of your stomach and intestines, eye movement, and equilibrium.

Inner View of the Brain

Cerebral cortex—Important for organization, memory, creativity, and its coverage of the cerebrum. The cerebrum has two hemispheres. The right side of the brain deals with spatial comprehension, musical abilities, and nonverbal ideas; the left side of the brain is important for language and analysis. Each hemisphere controls opposite sides of the body.

Corpus callosum—Acts as a bridge connecting the right and left hemispheres (parts) of the cerebrum.

Thalamus—Functions as a distribution center and transports incoming information from all parts of the nervous system. It is also important in taste, smell, pain, and memory.

Hypothalamus—Functions in temperature, sleep, hunger, thirst, pain, vomiting, and control of the pituitary gland.

Pituitary—Often called the master gland, it is vital in growth, maturation, and reproduction.

Brain stem—Handles basic functions, such as breathing, connects the cerebrum with the spinal cord, and is developmentally the oldest part of the brain. It consists of the midbrain, pons, and medulla oblongata.

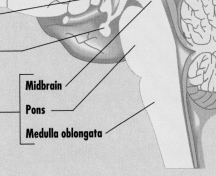

Midbrain

Pons

Medulla oblongata

Cerebellum—Involved in balance, and makes sure that movements are carried out correctly.

The Cranial Nerves of the Brain

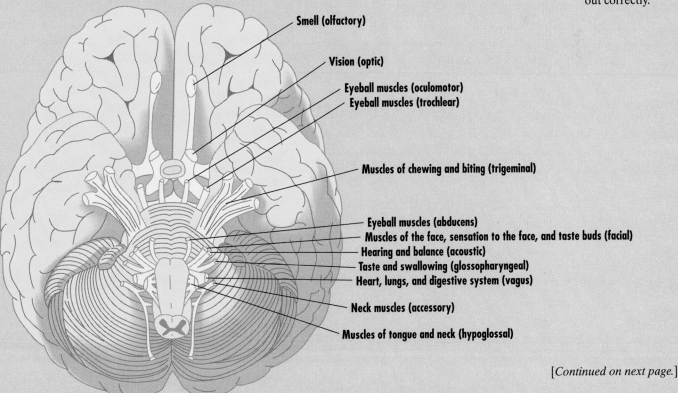

Smell (olfactory)

Vision (optic)

Eyeball muscles (oculomotor)
Eyeball muscles (trochlear)

Muscles of chewing and biting (trigeminal)

Eyeball muscles (abducens)
Muscles of the face, sensation to the face, and taste buds (facial)
Hearing and balance (acoustic)
Taste and swallowing (glossopharyngeal)
Heart, lungs, and digestive system (vagus)

Neck muscles (accessory)

Muscles of tongue and neck (hypoglossal)

[*Continued on next page.*]

Inside the Brain and Nervous System

Control Areas of the Cerebral Cortex of the Brain

The cerebral cortex is the largest portion of the brain. Specific areas in the cerebral cortex control a variety of functions of the body, ranging from motor coordination to speech and intelligence.

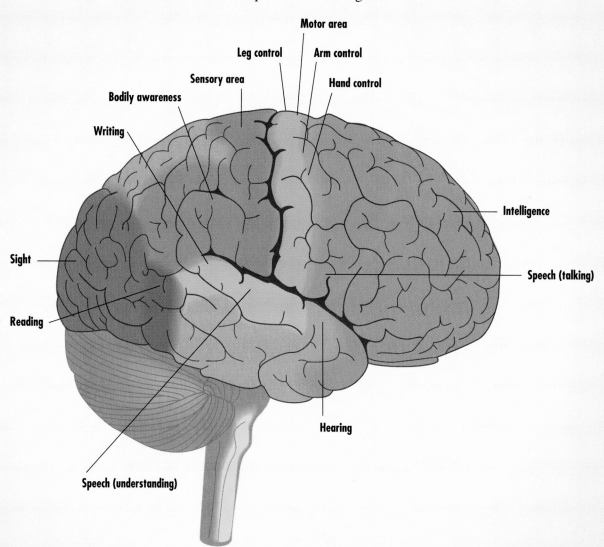

The Sensory Part of the Cerebral Cortex of the Brain

This figure shows in more detail the sensory control areas for different body parts on the sensory area of the cerebral cortex of the brain. Some areas of the body are represented by large areas on the sensory part of the cerebral cortex of the brain, while other parts have a small representation. The lips have the greatest area, followed by the face and thumb, with smaller amounts for the trunk and lower parts of the body. This means that the lips have the largest number of specialized nerves and are the most sensitive.

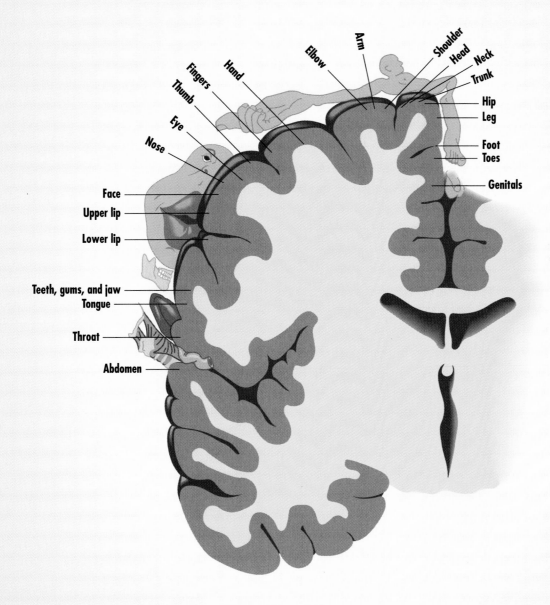

How the Brain Regulates Breathing and Pain

Breathing

Breathing is primarily regulated by the respiratory center in the medulla oblongata, and occurs approximately 10–15 times per minute. The respiratory center contains inspiratory (for inhalation) and expiratory (for exhalation) nerve cells. Additional receptors in the carotid arteries of the neck and in the aorta in the chest play an accessory role in regulating breathing. The medulla oblongata also communicates with another portion of the brain stem, the pons, which produces the rhythm of breathing in and out.

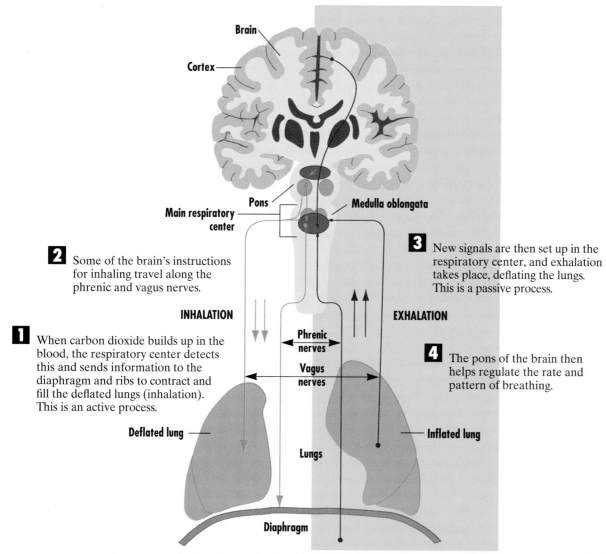

Brain

Cortex

Pons

Main respiratory center

Medulla oblongata

2 Some of the brain's instructions for inhaling travel along the phrenic and vagus nerves.

INHALATION

1 When carbon dioxide builds up in the blood, the respiratory center detects this and sends information to the diaphragm and ribs to contract and fill the deflated lungs (inhalation). This is an active process.

3 New signals are then set up in the respiratory center, and exhalation takes place, deflating the lungs. This is a passive process.

EXHALATION

Phrenic nerves

Vagus nerves

4 The pons of the brain then helps regulate the rate and pattern of breathing.

Deflated lung

Inflated lung

Lungs

Diaphragm

Pain

The brain also regulates pain. Skin tissues contain pain receptors, which respond to stimuli such as cutting or inflammation that can damage the tissue. A message travels from the receptors to the brain, which enables you to feel pain and then instructs the muscles to move the endangered tissue from the pain source.

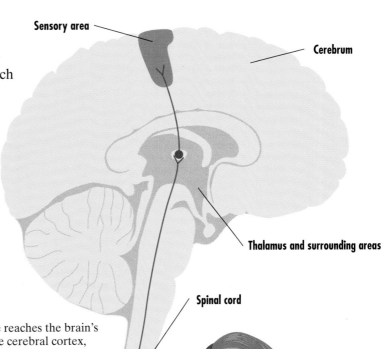

Sensory area

Cerebrum

Thalamus and surrounding areas

Spinal cord

4 When the impulse reaches the brain's sensory area in the cerebral cortex, the pain then attains a type of quality to it, such as sharp, electric, burning, or throbbing.

3 The impulse goes to the thalamus and nearby areas which produces pain.

2 This release of chemicals triggers a nerve impulse that travels to the spinal cord.

1 When you step on a tack, your injured tissues release chemicals that excite nerve endings.

Skin

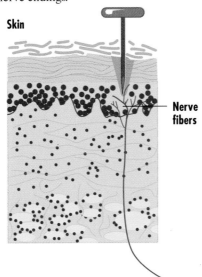

Nerve fibers

5 To prevent further injury to the tissue, the brain directs an impulse to the leg muscles, instructing them to lift the foot from the pain source.

When Something Goes Wrong with the Brain

Headaches

Tension Headaches   These headaches start as a tight band or viselike feeling in the muscles of the back of the head, and spread over the top of the head. The pain can last 30 minutes to 7 days or longer, and the majority of those who have these headaches are women.

Pain often in forehead, temples, or back of head and neck

Pressure on muscle may increase pain.

Trigger factors: muscle spasm, neck arthritis, or jaw/joint pain

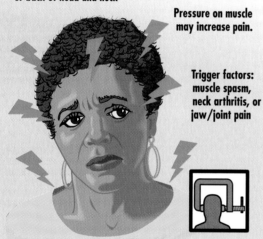

Vertigo

Visual auras

Confusion

Poor memory

May have local redness

Disturbed by sounds

Eyes irritated by light

Pale, perspiration

Thick speech

Nausea, vomiting

Migraine Headaches   Generally, these headaches produce a pulsating, hammerlike pain, which starts in the temple or the eye and can spread over the whole side of the head. The pain lasts 4 to 25 hours or longer, and again, the majority of those troubled with migraines are women.

Swelling and redness of eyelid

Severe headache behind eye

Pupils constricted and eyes tearing

Flushing of face and sweating

Nose congestion and secretion

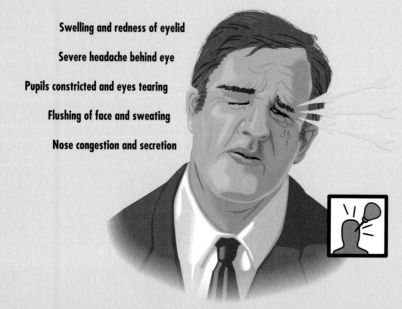

Cluster Headaches   These headaches appear as a boring, knifelike, excruciating severe pain, usually behind one eye. The pain lasts 30 to 90 minutes or more, and can occur in clusters for weeks to months. Typically, they affect large and strong men.

Strokes

Strokes can be caused by a blood clot in any portion of the brain. This stops the blood supply and creates damage. The damage may be mild or severe, permanent, long-lasting, or temporary. A person with paralysis of the right side of the body has damage on the left side of the brain. A person with a left-sided paralysis has the opposite. In addition to these symptoms, there can be dizziness, unsteadiness, clumsiness of an arm or leg, difficulty swallowing, and thinking or breathing problems.

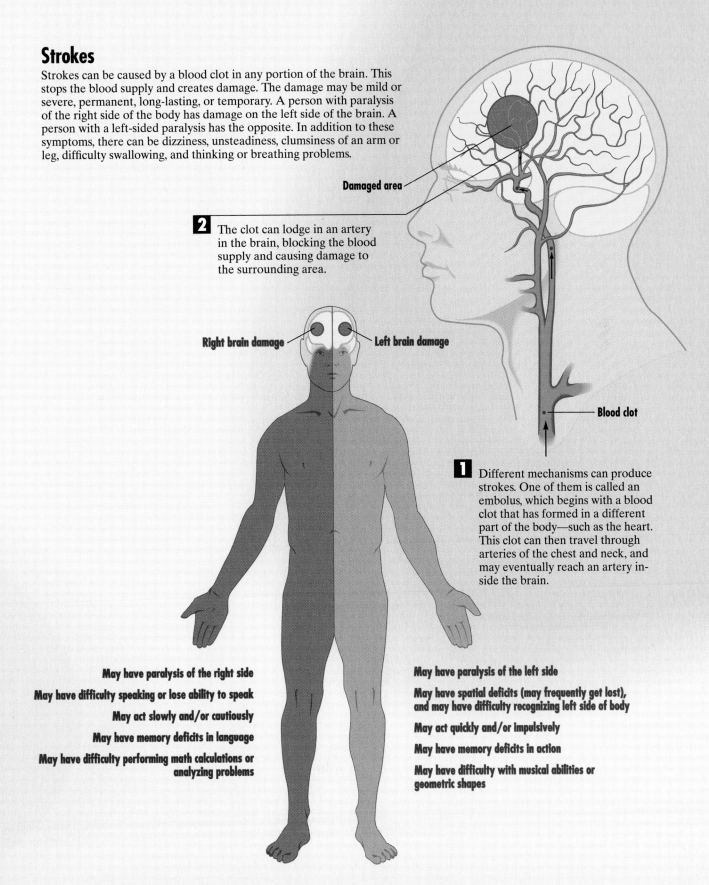

Damaged area

2 The clot can lodge in an artery in the brain, blocking the blood supply and causing damage to the surrounding area.

Right brain damage **Left brain damage**

Blood clot

1 Different mechanisms can produce strokes. One of them is called an embolus, which begins with a blood clot that has formed in a different part of the body—such as the heart. This clot can then travel through arteries of the chest and neck, and may eventually reach an artery in-side the brain.

May have paralysis of the right side

May have difficulty speaking or lose ability to speak

May act slowly and/or cautiously

May have memory deficits in language

May have difficulty performing math calculations or analyzing problems

May have paralysis of the left side

May have spatial deficits (may frequently get lost), and may have difficulty recognizing left side of body

May act quickly and/or impulsively

May have memory deficits in action

May have difficulty with musical abilities or geometric shapes

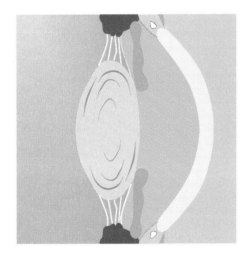

The Eye

THE EYE, nearly a sphere, may be considered one of the most useful approximate inches (by diameter) on the body. Its main purpose is to allow you to see and perceive the environment around you. This is accomplished in two ways. Objects transmit light to the light sensitive part of the eye, the retina, and then this information is translated by the visual part of the brain.

On the exterior surface, the eye's eyelid protects it and shuts out the environment. Deeper within, each eye has six muscles that rotate the eye horizontally, vertically, and obliquely. When the two eyes are tracking a moving object, the muscles of both eyes act together to keep the object on certain points of their respective retinas. If, for some reason, the coordination between the two eyes is lost, the images no longer fall on corresponding points, and double vision occurs.

Blinking is a reflex to actual or possible insults and also moisturizes the outer surface of the eye. It occurs by contraction of muscles in the upper and lower eyelids. Blinking spreads a protective film of lacrimal (tear) secretions over the surface of the eye and occurs every few seconds. However, reflex blinking can occur as fast as once every 0.1 second.

The body of the eye has three coats. The outer layer is the sclera, or the white of your eye. The front one-sixth of the sclera is called the cornea. Both the cornea and the conjunctiva (inner eyelids) are shielded by a fluid film of tears—secreted by lacrimal glands—and mucous and oily secretions. Tears are essential to wash away irritating particles and fumes. Secretion of tears also occurs with irritation of the cornea, the conjunctiva, or the nose, and with hot peppery foods, vomiting, coughing, and yawning.

The middle layer of the eyeball, the choroid, has an abundance of blood vessels that are important for nourishment to the eye. The front portion of this layer is composed of the iris (the colored portion of your eye) and the ciliary body (which suspends the lens of the eye and contains muscle fibers that help in vision). The degree of pigments determine the color of the iris. The lens, which is located centrally in this area, is composed of a strong elastic capsule filled with transparent fibers. When the lens is in a relaxed state with no tension, it assumes a spherical shape.

Between the cornea and the iris is the anterior, or front, chamber of the eye. This chamber, and the smaller portion behind it, are filled with a clear fluid, the aqueous humor. Behind this front chamber is the main body of the eye, which contains a transparent jelly—the vitreous humor. The aqueous and vitreous humors are important to maintain a normal intraocular (eye) pressure of generally 20 millimeters or less (the figure varies depending upon the instrument used to measure it). In order for this intraocular pressure to remain normal, the fluid is periodically and automatically drained away. If the pressure becomes abnormally high, then glaucoma, an eye disease, has developed.

The major part of the inner layer of the eye, the retina, is very important because it transmits visual information to the brain and contains cells called cones and rods. Cones control color vision (with photo pigments of red, blue, and yellow) and visual acuity, and rods are important for night vision, light detection, and movement. When a single group of color-receptive cones is missing from the eye, you are unable to distinguish some colors from others. If red cones are missing, you cannot differentiate the spectrum of colors from green through red, with all of these colors appearing as green. If green-sensitive cones are missing, you sense only red in the color range of green through red. If there is a loss of blue receptors, you have a greater quantity of green, yellow, orange, and red colors in your visual spectrum and may not see blue colors well, or at all. (A rapid method for determining color blindness is based on spot charts.) Rods contain only one pigment, called rhodopsin.

The cells of the retina are also important for projecting visual information to the optic nerve of the eye, and then to the brain. Acuity is the eye's response to a visual image and ultimately determines how well you see. Vision charts are used by ophthalmologists (eye physicians) and optometrists to determine the extent of your visual acuity; 20/20 vision is considered perfect. If you have 20/20 vision, it means that you are able to see clearly at 20 feet what a normal eye can see at 20 feet. Other potential visual abnormalities can involve visual fields, color vision (described above), and depth perception.

A visual field is the area seen by an eye and appears on the retina. The amount of this field is limited by the nose and the eyebrow ridges from above and by the structure of the retina. There are no vision receptors in the area where the optic nerve leaves the retina (to go to the brain). Therefore, if an image falls in this location, it is not seen. This is called the blind spot. Still another component of vision is depth perception, which is the ability of the eyes to perceive distance.

As you age, so does your eye. This can cause significant impairment in some people. First, there are decreases in visual acuity as a result of changes in the retina, lens, or nerves, and changes in the refractive power of the eye which can lead to nearsightedness or farsightedness. Also, there often is decreased tear secretion, which produces irritation and discomfort. Although there are some people who continue to enjoy perfect eyesight throughout all or most of their lives, most of us sooner or later develop what is known as nearsightedness or farsightedness.

About 30% of Americans are nearsighted, or myopic. This common disorder prevents one from seeing faraway objects clearly, but the vision of near objects is not affected. These distant objects are blurred, because either the eyeball is too long or the lens of the eye is too curved. Myopia is corrected by the use of glasses or contact lenses. Approximately 60% of Americans are farsighted, or hyperopic. This means that faraway objects are seen well, but close objects are blurred. Farsightedness occurs because the eyeball is too short, or the lens is too weak. It can also be corrected with glasses or contact lenses.

Although there are many eye diseases, we will focus on only a few. The first is glaucoma, which can produce a slow- or fast-rising increase in pressure within the eye. Pressure of the aqueous humor fluid in front of the lens increases pressure in the vitreous humor behind the eye. When the flow of the aqueous humor fluid, within the eye and traveling out of the eye, is decreased, resulting in increased eye pressure, damage can occur to the rods and cones, and there can be destruction of the optic nerve. Blindness can occur quickly if treatment is not provided.

Another common eye condition is the development of cataracts, and it is estimated that 3.6 million Americans have them. They appear as a film on the lens of the eye, which causes a gradual and painless deterioration of sight.

Let us now explore the eye—often called a mirror to your soul—which does act as a mirror to reflect into the body the images of the world around you.

How You See

The eye collects light from objects and projects them on the light-sensitive portion of the eye, the retina. This information is then translated by the visual part of the brain. Color images are formed from both eyes simultaneously, providing a three-dimensional image.

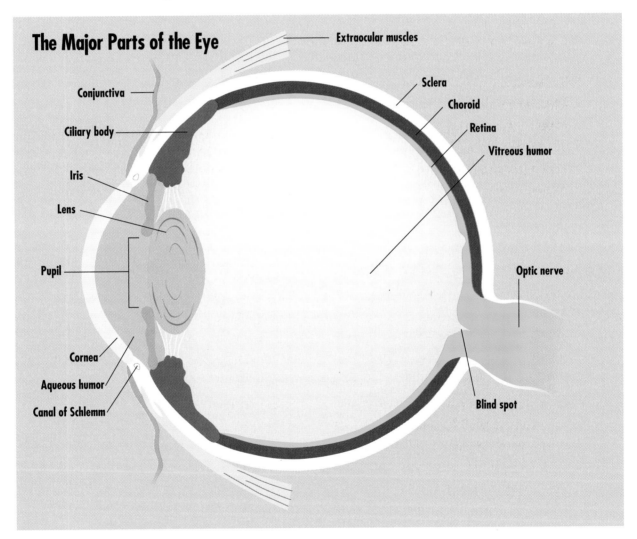

The Major Parts of the Eye

Extraocular muscles

Conjunctiva

Ciliary body

Iris

Lens

Pupil

Cornea

Aqueous humor

Canal of Schlemm

Sclera

Choroid

Retina

Vitreous humor

Optic nerve

Blind spot

As light enters the eye, it passes through the cornea, into the aqueous humor, then to the pupil and lens. The lens of the eye changes its shape in order to focus the light as it further enters the eye. From there, light passes to, and is focused on, the retina; this process is called refraction. Following that, the impulses go to the optic nerve and then to the brain for interpretation.

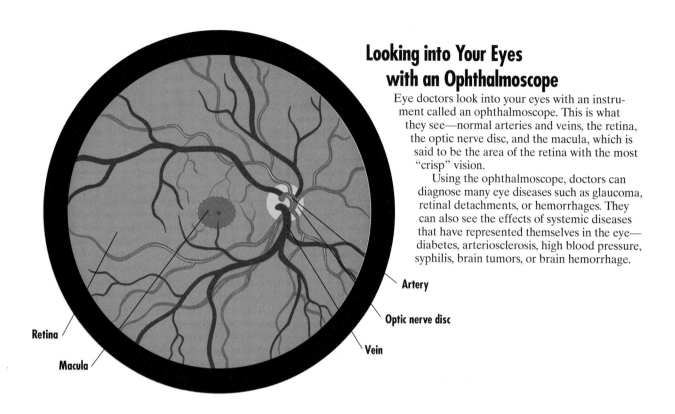

Looking into Your Eyes with an Ophthalmoscope

Eye doctors look into your eyes with an instrument called an ophthalmoscope. This is what they see—normal arteries and veins, the retina, the optic nerve disc, and the macula, which is said to be the area of the retina with the most "crisp" vision.

Using the ophthalmoscope, doctors can diagnose many eye diseases such as glaucoma, retinal detachments, or hemorrhages. They can also see the effects of systemic diseases that have represented themselves in the eye—diabetes, arteriosclerosis, high blood pressure, syphilis, brain tumors, or brain hemorrhage.

Artery

Optic nerve disc

Vein

Retina

Macula

How the Eye Focuses on Objects

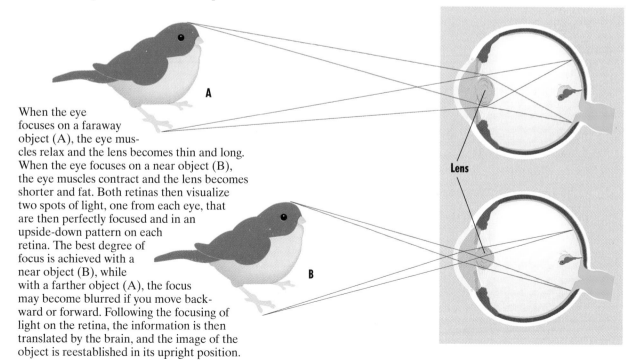

When the eye focuses on a faraway object (A), the eye muscles relax and the lens becomes thin and long. When the eye focuses on a near object (B), the eye muscles contract and the lens becomes shorter and fat. Both retinas then visualize two spots of light, one from each eye, that are then perfectly focused and in an upside-down pattern on each retina. The best degree of focus is achieved with a near object (B), while with a farther object (A), the focus may become blurred if you move backward or forward. Following the focusing of light on the retina, the information is then translated by the brain, and the image of the object is reestablished in its upright position.

A

B

Lens

Nearsightedness and Farsightedness

Nearsightedness (difficulty seeing far objects) and farsightedness (difficulty seeing near objects) can be corrected with eyeglasses. The strength of the proper lens used in eyeglasses is determined by trial and error during an optical examination. First a strong lens is tried, then a weaker lens, and so on, until the lens producing the best vision for you is determined.

Normal eye

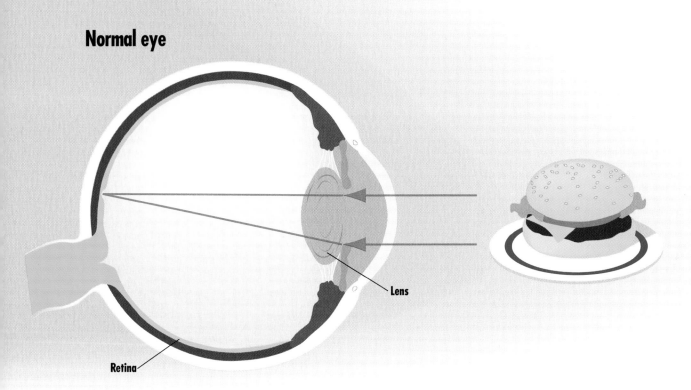

Lens

Retina

Nearsightedness (Myopia)

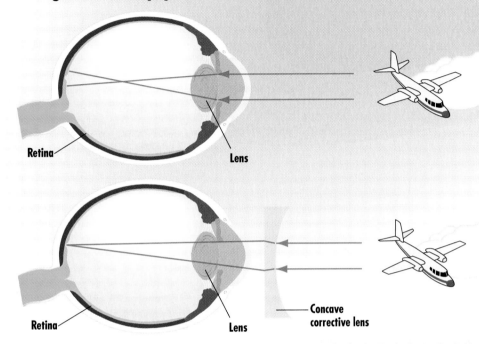

Retina   Lens

The lens in the eye of a person with nearsightedness often has too much refraction, and/or the eyeball itself is too long. This defect produces the image not on the retina itself, but in front of it, resulting in unclear vision.

Retina   Lens

Concave
corrective lens

Nearsightedness is corrected with eyeglasses with concave lenses. With corrective lenses, the image is produced on the retina, not in front of it.

Farsightedness (Hyperopia)

A person with farsightedness has too short an eyeball or too weak a lens. This defect focuses the image beyond the retina, not on it as in the normal eye.

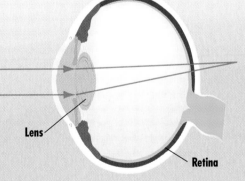

Lens

Retina

Farsightedness is corrected with eyeglasses with convex lenses. With corrective lenses, the image is focused on the retina, not beyond it.

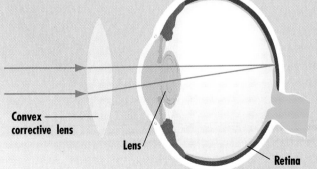

Convex
corrective lens

Lens

Retina

Glaucoma

In glaucoma, the aqueous humor fluid has difficulty flowing through the pupil into the front chamber and also in leaving the eye via the canal of Schlemm. This produces excess pressure build up, which can cause damage or blindness.

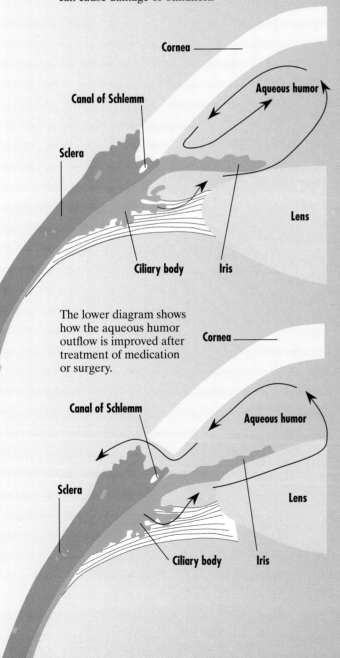

Cornea

Aqueous humor

Canal of Schlemm

Sclera

Lens

Ciliary body   Iris

The lower diagram shows how the aqueous humor outflow is improved after treatment of medication or surgery.

Cornea

Canal of Schlemm

Aqueous humor

Sclera

Lens

Ciliary body   Iris

Primary glaucoma usually occurs without any definite cause. Certain families appear to have a predisposition for this, and it may be more common in people who are far-sighted. Secondary glaucoma is usually a complication of other eye disorders, such as iritis, and people with diabetes or recently controlled high blood pressure can frequently develop this. Glaucoma can occur slowly with minimal or no symptoms, or it can occur within hours, and the person may have severe pain, nausea and/or vomiting, see rainbow-colored halos around lights, and have blurred and/or lost vision. The vision may be permanently lost if proper treatment is delayed.

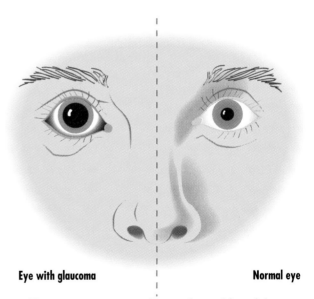

Eye with glaucoma   **Normal eye**

Glaucoma can cause swelling and stretching of the cornea, as above, so that it can appear very large, and advanced stages can produce blue-gray opacities of the cornea. The pupil of the eye may be enlarged (dilated), will not react to light, and the cornea may be steamy looking with fluid (edema). The diagnosis can be confirmed by using an instrument called a tonometer, or slit lamp instrument, which measures the pressure of the eye.

Cataracts

Cataracts usually develop slowly, without pain, redness, or tearing. They can result from old age, heredity, infection from diseases like German measles early in life, diabetes, eye injuries, or the use of drugs like steroids. The usual symptoms include blurred vision, a feeling of a film over the eye, perceived changes in the color of objects, and/or a need for frequent eyeglass changes.

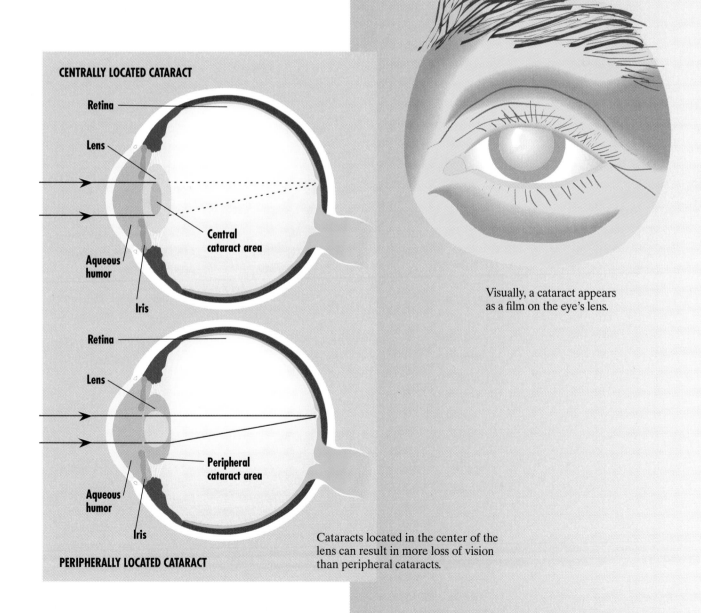

CENTRALLY LOCATED CATARACT

Retina

Lens

Central
cataract area

Aqueous
humor

Iris

Retina

Lens

Peripheral
cataract area

Aqueous
humor

Iris

PERIPHERALLY LOCATED CATARACT

Visually, a cataract appears as a film on the eye's lens.

Cataracts located in the center of the lens can result in more loss of vision than peripheral cataracts.

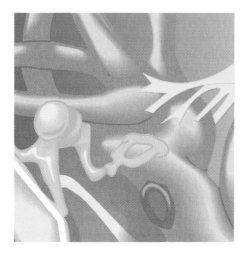

The Ear

YOUR EAR IS an exquisite instrument with a remarkable design. One of its functions is to hear sounds. It then transmits the sound as impulses to the brain, which interprets and reacts to them. Another important function of your ear is to maintain balance and equilibrium.

Sound is made by nondetectable vibrations of objects (alive and inanimate) that produce a pattern of alternating compression (increased pressure) and decompression (decreased pressure) of the air. The quality of the sound that you hear depends on a combination of frequencies. Music or pleasant sounds consist of one or more related frequencies. Noise, however, consists of a mixture of unrelated frequencies. The intensity of a sound is measured in decibels: the larger the number of decibels a sound produces, the louder the sound is to the ear.

When sound enters your ear and hits the eardrum (tympanic membrane), it puts the eardrum into motion. The eardrum is attached to the smallest bones in your body, the ossicles. They are located in the middle ear and send sounds to the cochlea of the inner ear, the snail-shell-type structure that is an essential organ of hearing. The fluid in the inner ear then causes another membrane, the basilar membrane of the inner ear, to move in a way determined by the frequency of sound. The movement of the basilar membrane stimulates other elements in the inner ear, so that messages are transmitted by the acoustic nerve to the proper area of the brain, the auditory cortex, where the sound is then interpreted.

The ear comprises three main components: the outer ear, middle ear, and inner ear. The outer or *external ear* consists of the funnel-shaped auricle (outside part of the ear), external ear canal (the passageway into your ear), and the cone-shaped tympanic membrane, which prevents objects from entering the ear and keeps the temperature of the eardrum stable. Hairs and wax (cerumen) in the ear canal further help to protect the ear.

The *middle ear*—which has three small ossicles or bones: the malleus, incus, and stapes—is important for allowing movement of the sound messages into the inner ear, to protect the structure of the inner ear from rough movements, and to improve hearing in a noisy environment. The middle ear is also protected by two muscles—the tensor tympani and the stapedius—which act to reduce the transmission of low-frequency sounds. Low-frequency sounds can be excessively loud and may damage the basilar membrane of the cochlea. These two muscles also function to decrease your sensitivity to the sound of your own speech.

The *inner ear* is essential for both hearing and balance control, and basically consists of the cochlea, semicircular canals, and two chambers—the saccule and utricle. The cochlea, which is responsible for your hearing capabilities, consists of coiled tubes filled with fluid and is separated by membranes. The vestibular apparatus of the inner ear keeps you from being dizzy, and consists of three semicircular canals that govern angular movements of the head, and the saccule and utricle. The horizontal canal governs vertical turning movements, such as dancing the waltz. The vertical canals regulate movements in the horizontal axis, such as leaping and landing after throwing a basketball through a hoop. Balance and hearing messages from the inner ear are then transferred to the brain, via the acoustic cranial nerve, where the brain interprets this information.

Seventy-six million Americans will suffer from an inner-ear disorder, including dizziness or vertigo, at some time in their life. The resulting loss of balance can be frightening: Dizziness is any sense of altered orientation in space (such as lightheadedness, giddiness, or unsteadiness), and vertigo is an illusion of motion (as if the world were revolving around you or you were revolving in space). Other accompanying symptoms for inner-ear disorders can include nausea, ear ringing, hearing loss, weakness, headache, pallor, and perspiration. Inner-ear damage can be caused by infections, injuries to the head, arteriosclerosis, allergies, or neurologic diseases. Tests that may be done to check the amount and type of damage include checking for nystagmus, which are abnormal eye movements, and measuring these eye movements with a test called an ENG (electronystagogram); a hearing test; blood tests; or a CT scan of the brain. Other tests could include platform posturography and/or rotation testing.

Deafness is another ear disease of large proportion. It has been estimated that in the United States alone, there are about twenty-two million people of all ages with significant hearing loss. There are two basic types of hearing difficulties. A problem with the cochlea or the auditory nerve, called *nerve deafness*, can result from the flu, meningitis, other infections, Ménière's disease, arteriosclerosis, certain drugs, or repetitively loud sounds. *Conduction deafness*, an impairment of the middle ear, results from ear wax, infections, genetic problems, or deep-sea diving.

Deafness is usually easy to detect in adults by using standard hearing tests. However, loss of hearing may be more difficult to detect in children. Initially, a child with hearing difficulties may act bored, uninterested, or have difficulty speaking. Often, more sophisticated tests need to be done in young children to measure hearing loss.

We shall now travel through a winding road into the ear and learn more about it.

The Hearing Pathway

Sound waves travel through the external ear, into the middle ear, and then to the inner ear. Once the sound waves reach the inner ear, they are transmitted to the cochlea, which consists of coiled snakelike tubes filled with fluid and separated by membranes. The cochlea contains a structure called the basilar membrane, on top of which lies the organ of Corti. The sound impulses are then transmitted to the basilar membrane, which causes the membrane to vibrate. These vibrations are then relayed to the hair cells within the organ of Corti, which then causes these hair cells to become excited. The total number of hair

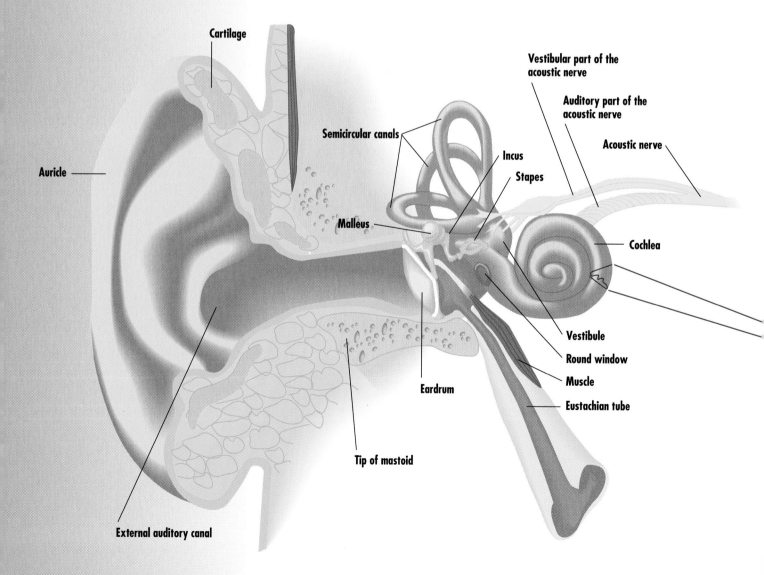

Cartilage

Vestibular part of the acoustic nerve

Auditory part of the acoustic nerve

Semicircular canals

Incus

Acoustic nerve

Stapes

Auricle

Malleus

Cochlea

Vestibule

Round window

Muscle

Eardrum

Eustachian tube

Tip of mastoid

External auditory canal

cells is approximately 20,000, and if damaged, they usually cannot be replaced. Outer hair cells may be important in determining the intensity of sound, and inner hair cells may be important for pitch. The loudness of the tone is also believed to be due to the amount of basilar membrane set into motion and the rate and number of hair cells that vibrate. Impulses from the hair cells are then sent to the acoustic nerve, then to the brain stem, and then to the cortex of the brain. In the brain, these impulses are translated into perceptions and detailed information is presented to you regarding the sound.

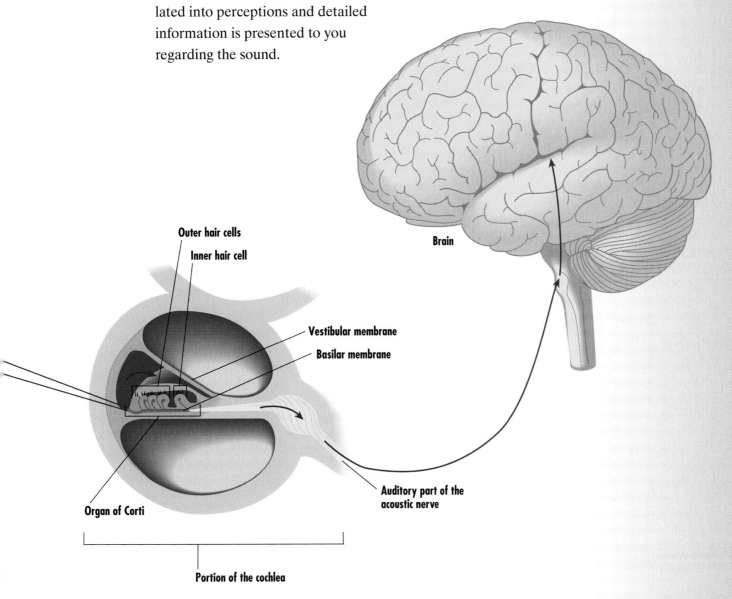

Outer hair cells

Inner hair cell

Brain

Vestibular membrane

Basilar membrane

Organ of Corti

Auditory part of the acoustic nerve

Portion of the cochlea

The Ear and Balance

The main coordinating center for balance is the brain stem of your brain. It processes information from your eyes, neck, muscles, joints, and the vestibular system (balance part) of your inner ear. The inner ear, or labyrinth, is a complicated maze of chambers in the skull. The information from the vestibular system in your inner ear provides critical data on your balance and where you are in relation to the environment.

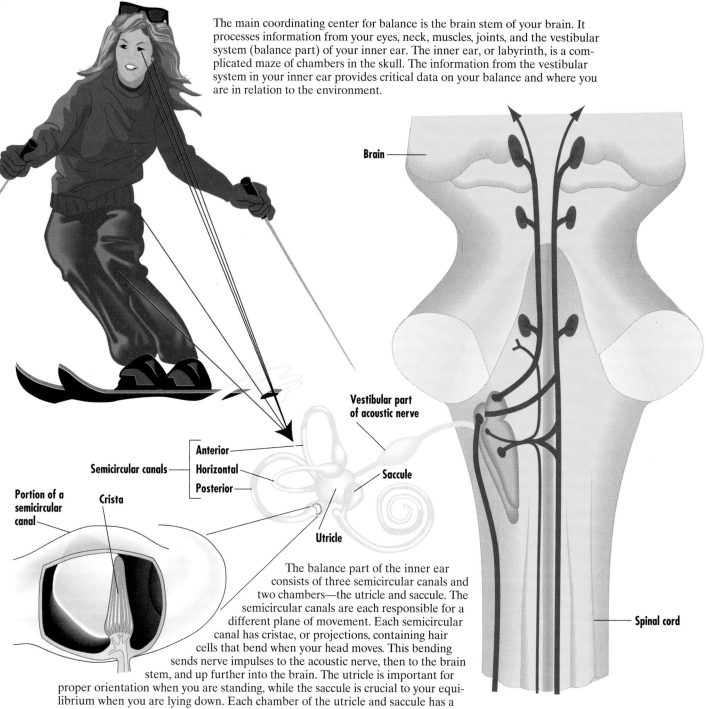

Brain

Vestibular part
of acoustic nerve

Anterior
Semicircular canals   Horizontal   Saccule
Posterior

Portion of a
semicircular   Crista
canal
Utricle

Spinal cord

The balance part of the inner ear consists of three semicircular canals and two chambers—the utricle and saccule. The semicircular canals are each responsible for a different plane of movement. Each semicircular canal has cristae, or projections, containing hair cells that bend when your head moves. This bending sends nerve impulses to the acoustic nerve, then to the brain stem, and up further into the brain. The utricle is important for proper orientation when you are standing, while the saccule is crucial to your equilibrium when you are lying down. Each chamber of the utricle and saccule has a *macula*, a layer of hair cells coated with a jellylike substance. Every macula also contains otoliths, small granules of calcium, and when the head moves, it also moves the otoliths and hair cells. Again, as a result of stimulation to the hair cells, nerve impulses are sent to the brain along the acoustic nerve. The cerebellum and brain stem are of utmost importance in equilibrium, with the brain providing instructions on how to correct the position of the body, thereby making it more stable.

Diagnosing Balance Disorders

When doctors need to find out the interrelationship between all parts of your balance system—the inner ear, muscles, joints, and eyes—they may use a test called platform posturography for a diagnosis. The patient stands on a computer-controlled floor, and electrodes can also be attached to the person to further monitor their movements. An enclosure with patterns of dots, clouds, or stars surrounds the patient, and a parachute-type harness protects the person from falling when the floor position and the enclosure change. If the vestibular system is not working well, the patient cannot maintain his or her balance. The results can then help the doctors determine the cause and degree of the problem.

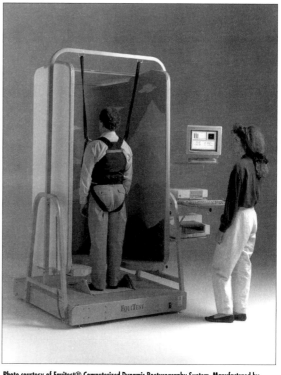

Photo courtesy of Equitest® Computerized Dynamic Posturography System, Manufactured by NeuroCom International, Inc., Clackamas, Oregon.

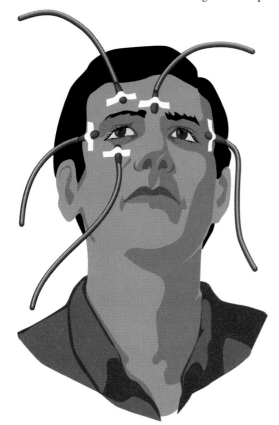

Dizziness and Abnormal Eye Movements

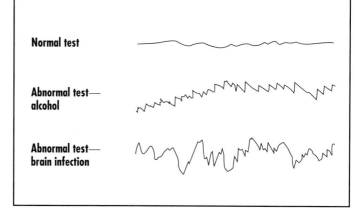

Abnormal eye movements (nystagmus) are common in people who have inner-ear disorders, vertigo, or dizziness. These eye movements are involuntary, and alternate between rapid and slow movements. An ENG (electronystagmogram) is a common test used for eye movement measurements, and in preparation for this test, electrodes are positioned on the face in specific spots. A technician also inserts water into each ear, at which time the electrodes record the eye movements that may help determine a diagnosis. This illustration shows how the electrodes for the ENG test are often placed on a person's face. The ENG test tracing depicted at the top is normal and is almost flat horizontally. This person had no dizziness or vestibular problem. The middle test result is abnormal and is from a person who drank an excessive amount of alcohol and was dizzy. This tracing shows nystagmus, and each beat of the eye is represented by one upward and downward movement of the tracing. The bottom tracing is also abnormal and is from a patient who had a severe brain infection with damage to the brain and the ear. Nystagmus is also present, but is more disorganized and severe, with no two successive eye movements, or beats, having anything in common with the other.

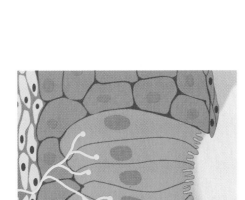

C H A P T E R
4

Your Nose and Senses
of Smell and Taste

THE MOST IMPORTANT function of your nose is to breathe in air, and of your tongue is to assist in the swallowing of food; however, their most enjoyable functions are smelling and tasting, respectively. Both the nose and the tongue respond to similiar stimuli, such as the aroma and flavor of a home-cooked meal, and they are often interdependent on each other. For example, the appreciation of that home-cooked meal is enhanced by the aromas permeating from the kitchen, which heightens the taste sensation, and this in turn can intensify the sensation of the aromas coming into the nose.

The purposes of your nose include warming the air, breathing in the air, filtering the air and its particles, and smelling. Anatomically, the nose has two portions, separated by an internal bony partition which is called the nasal septum. The nose also has a number of blood vessels that help to warm the inhaled air. The mucous glands of the nose make sure that the moisture content of the air, which eventually reaches the lungs, is close to 100% and that large particles are trapped and moved out and away, to be coughed up or swallowed. Over 95% of these particles are filtered out, with the hairs at the entrance of the nose removing larger particles, and the adenoids (lymph gland tissue in the nose-throat area) at the exit, filtering smaller particles and organisms. Additionally, your nose defends against infection by moving bacteria and viruses out of the nose, and its mucus also contains antibodies to neutralize these viruses and bacteria. A remarkable one quart of mucus is produced per day!

Another function of the nose, smell, is not as acute as some of our other senses, such as seeing or hearing, and it is poorly developed compared to that of "man's best friend," the dog. In order for you to smell a substance, it must be volatile (so that it can pass through mucus) and it must also be water- and fat-soluble. There are more than 50 different types of olfactory stimulants, although only seven have been defined: camphor, musk, floral, and peppermint, as well as ethereal, pungent, and putrid. The acuity of our sense of smell, just like our hearing and seeing, decreases with age.

Not that much is known about how our sense of smell functions. A current theory is that specified odors attach to nose cells and that causes them to become excited. This excitation causes messages to be relayed to the olfactory nerve and then to the brain. There are extensive connections between the olfactory (smelling) apparatus and the hypothalamus, brain stem, and other parts of your brain. Doctors and researchers are learning that your sense of smell may be even more important than previously recognized; for instance, certain aromas can bring back past memories of complicated events, which you would not have otherwise remembered. Also, certain smells appear to be involved in the control of your food intake and whether you feel awake, happy, or sleepy. Your sense of smell is most often affected by colds, flu, or allergies. It can also be impaired by head injuries, meningitis or other infections, and Alzheimer's disease.

Another important and enjoyable sense organ in your body is your tongue, which contributes to your sense of taste. Your sense of taste is crucial for the enjoyment of food and drink and to help recognize poisons. Without taste, you could not enjoy an aromatic cup of coffee, a tangy piece of deep-dish pizza, or a sinfully rich hot chocolate sundae. Four basic tastes are determined by your taste buds, which are found on the front, back, and sides of your tongue: sweet, bitter, sour, and salty. Bitterness is the most sensitive taste, followed by sourness, saltiness, and then sweetness. Although all four tastes can be perceived at the tip of the tongue, the tip is most sensitive to sweetness and saltiness. The sides of the tongue are most sensitive to sourness, and the back of the tongue to bitterness. We assume that you can taste more than these four basic sensations, but these are combinations, and involve not only taste, but smell. Taste is just one component of what we call flavor, and flavor involves aromas that travel up the back of the throat to the olfactory (smelling) receptors in your nose. Further research is being conducted in this area.

Taste is determined by taste buds, the specialized nerve endings found on the tongue. You have about 10,000 of these taste buds. Each one is only about 1/30th of a millimeter in size, lives only about ten days, and is then replaced. Unfortunately, the numbers of taste buds also decrease with age. Each taste bud responds primarily to one type of taste, and once a taste is detected, the taste buds send messages to one or more of the cranial nerves, then to the thalamus, and finally to the cortex of the brain for further interpretation.

The most common cause of taste abnormalities is heavy smoking and, in particular, pipe smoking. Impairment can also result from a flu or a cold, dryness of the mouth, deficiencies of vitamins B_{12} or A, cancer, drugs, possibly low zinc, and some unknown reasons.

Now, it is time for us to learn more about how you can smell a sweet red rose, or taste some creamy mint chocolate chip ice cream.

The Olfactory (Smelling) System

Aromas or odors are received by the olfactory (or smelling) nerve cells of the nose. These tapered cells, along with other components, make up filaments that pierce the cribriform plate of the skull—a bony sievelike plate that allows the olfactory nerve filaments to combine and enter the olfactory bulb. From there, a nerve plexus is formed, which sends messages to the brain—the cerebrum, hypothalamus, and brain stem. These olfactory pathways in the brain provide for the control of your food intake, your dislike of the smell of certain foods, and your conscious perception of smells.

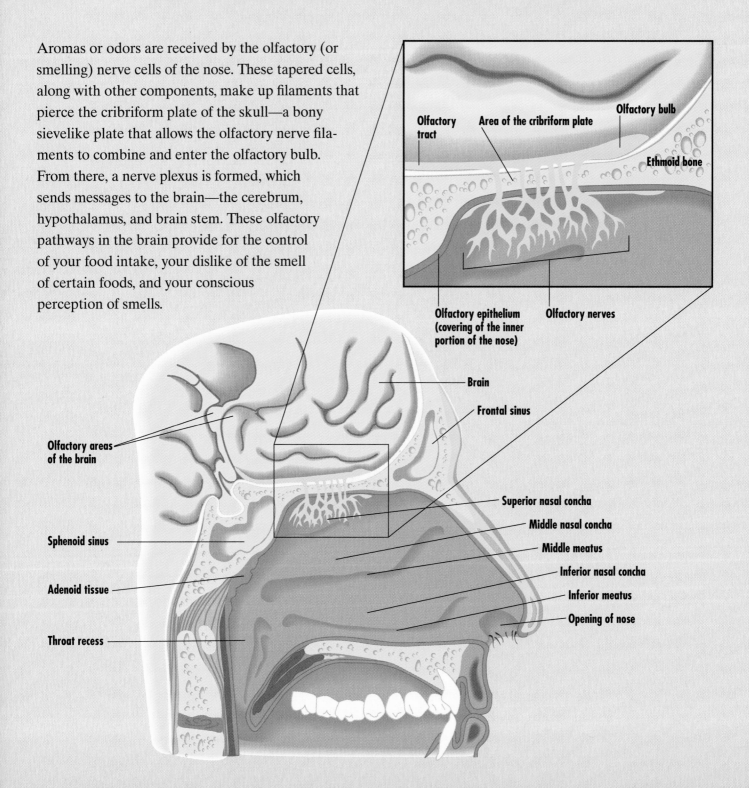

Olfactory tract

Area of the cribriform plate

Olfactory bulb

Ethmoid bone

Olfactory epithelium (covering of the inner portion of the nose)

Olfactory nerves

Brain

Frontal sinus

Olfactory areas of the brain

Superior nasal concha

Middle nasal concha

Middle meatus

Inferior nasal concha

Sphenoid sinus

Inferior meatus

Opening of nose

Adenoid tissue

Throat recess

The Sinuses

The sinuses, which lie above your nose, may function in adding resonance to the voice, producing small amounts of mucous, warming and humidifying air, and forming a type of shock absorber. Recent research has shown that even the common cold can produce sinus obstruction in most of us.

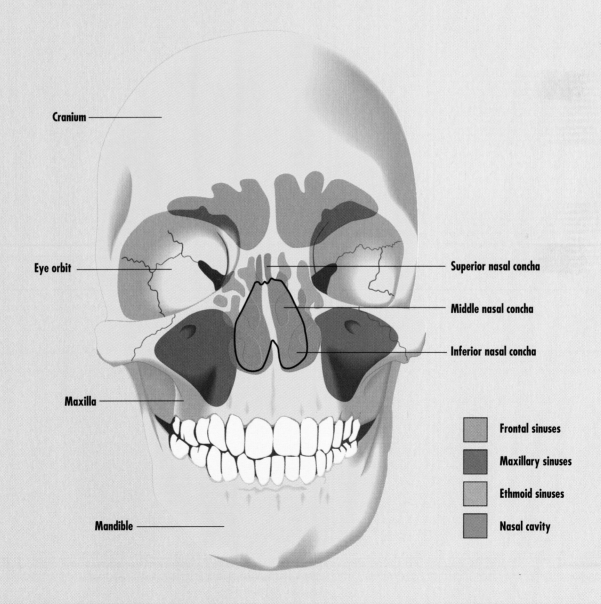

Cranium

Eye orbit

Maxilla

Mandible

Superior nasal concha

Middle nasal concha

Inferior nasal concha

Frontal sinuses

Maxillary sinuses

Ethmoid sinuses

Nasal cavity

How You Know if It's a Pickle or Ice Cream

The Pathway of Taste

Many taste buds are found on the tip, sides, and body of the tongue, although other taste buds, specifically the circumvallate papillae, are located primarily on the back of the tongue, and to a lesser degree on the pharynx (throat), and larynx (voice box). Taste buds are supplied by the facial (entered into by the trigeminal nerve), glossopharyngeal, and vagus cranial nerves, which carry taste messages to the thalamus and then to the cortex of the brain for interpretation. When you eat a pickle, salt taste buds on the tip and sour taste buds on the sides of your tongue are activated and send nerve impulses to your brain. If you eat ice cream, the taste buds for sweetness, which are also found on the tip of your tongue, become excited, and let your brain know how to react to experiencing this taste.

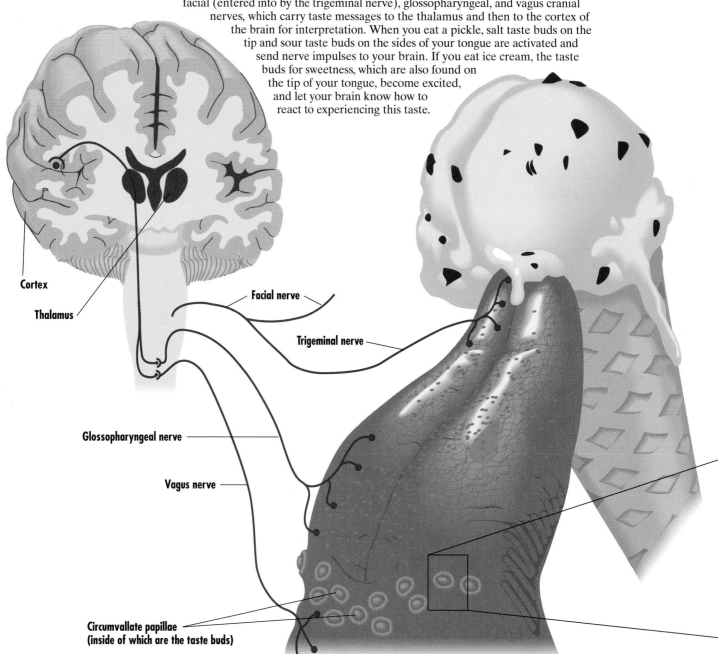

Cortex

Thalamus

Facial nerve

Trigeminal nerve

Glossopharyngeal nerve

Vagus nerve

Circumvallate papillae
(inside of which are the taste buds)

Microscopic Representation of a Taste Bud

Each taste bud comprises approximately 40 taste cells. The outer tips of the taste cells surround a taste pore, and the tip of each cell has protruding taste hairs (microvilli) that point up and out of the tongue. It is believed that these hairs provide the main receptor surface for taste.

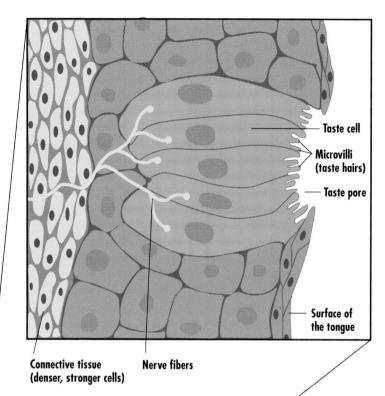

Taste cell

Microvilli (taste hairs)

Taste pore

Surface of the tongue

Connective tissue (denser, stronger cells)

Nerve fibers

Circumvallate papilla

Surface of the tongue

Circular trough

Taste bud

Mucous gland

Lymph nodule

Muscle fibers

Part of the Tongue Showing Taste Buds in a Circular Trough

Taste buds are situated on most of the different types of papillae (projections) found on the tongue. However, they predominantly occur on the walls of the circular troughs of the *circumvallate papillae*, which are mainly located on the back of the tongue.

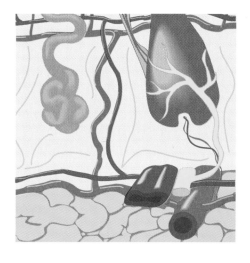

CHAPTER 5

The Skin, Hair, and Nails

YOUR SKIN COVERS the entire surface of your body, is the body's largest organ, and is essential for life. The offshoots of the skin are the hair, nails, and sebaceous/sweat and other skin glands. You have skin to provide you with a waterproof covering; protect you from toxins, damage, and infection; maintain the temperature of your body; and eliminate certain wastes. Your skin is the most extensive sense organ of the body; it receives stimuli such as touch, pressure, pain, and temperature, and it repairs itself. It is also important in your body's production of vitamin D from ultraviolet sunlight.

The skin has three basic layers—the outer epidermis, the inner dermis, and the more inner subcutaneous layer.

The epidermis itself has two layers—an outer covering of dead cells and keratin, the stratum corneum, and an inner covering, the malphigian layer, which contains melanin and keratin. Melanin is important for skin color. The darkness of your skin depends on the number and types of melanocytes or melanin cells, which are determined by genetics, the amount of sun exposure, and hormonal factors. Additionally, your skin color may vary with temperature, exercise, emotions, and your general health. Africans and African Americans may have a slightly higher number of melanocytes than do whites, but their difference in color is due more to the quality of the melanocytes and the way in which the pigment is scattered. The melanin concentration in whites is primarily in the deeper areas of the malphigian layer, whereas in blacks the granules of melanin are spread throughout the entire malphigian layer. Keratin is a protein found in the skin, and also found in the hair, nails, and teeth. It provides strength, makes the skin waterproof, and when there has been excessive rubbing or pressure on the skin, it forms a callus.

The dermis consists of hair follicles, blood vessels, nerves, muscle tissue, and sweat, oil, and other glands. All of these components are loosely connected by collagen, a protein substance found in the skin and also in other tissues such as tendons, bones, and cartilage.

The subcutaneous layer of the skin is a network of bundles of fascia (bands of fibrous tissue deep in the skin), connective tissue, and fat cells. This layer helps permit the movement of the skin. Children have an abundance of this subcutaneous layer, which helps prevent heat loss and cushions their body from injury during the rough-and-tumble years. As we age, this baby fat becomes redistributed. In women, it goes to the breasts, hips, thighs, and buttocks. In men, it goes to the neck, lower back, upper arms, and backside.

The hair functions to protect areas such as the eyes, ears, and nose from dust, insects, and other matter. Hair is found on all parts of the skin except the soles, palms, umbilicus, certain parts of the genitalia, and orifices. Hair consists of a root originating in the dermis and an elastic, horny filament that may grow out of the skin to a length of five feet or more, and is made up of keratin, a protein that grows in the hair follicle. Each hair has three layers, with color forming in the middle layer. Associated with hair are glands and muscle fibers. These glands lubricate the hair, and the muscle helps to extrude the lubricant from the gland. All hair eventually falls out, as a part of cycles of growth and hair loss.

Fingernails and toenails are found only in humans and primates and are a specialization of the epidermis. Fingernails grow about four times faster than toenails, and the rate is quicker in the summer months than in the winter.

The glands of the skin are important and varied, and include the sweat, sebaceous, lacrimal, and mammary glands. (The sweat and sebaceous glands are described further in the figure titled "The Skin, Hair, Nails, and How Your Skin Repairs Itself," which appears later in this chapter.) Lacrimal glands are found on the conjunctiva of the eyes. The secretions they form are more commonly known as tears—clear and salty, they moisten the cornea and the surrounding area. The mammary glands are found in adult females and produce milk. (Mammary glands are further discussed in Chapter 15.)

There are a great number of variations and abnormalities of the skin. The most important, and one of the most potentially serious, is skin cancer, but many other skin conditions are usually benign. As we age, our skin changes, and everyone develops wrinkles. We were all born with skin creases and lines, and as we grow older the number of these creases and lines does not change, but the physical structure of the skin does. The skin's elasticity, amount of collagen, and strength all decrease. This causes skin atrophy or thinness, which makes these creases and lines more obvious—now they're called wrinkles. Freckles are also very common, harmless lesions found predominantly in children. They are the result of melanocytes being grouped together in a clump or clumps. These light brown spots, with irregular borders, usually become more prominent or appear after sun exposure, and usually fade later in the fall or winter. The sunlight activates these clumped melanocytes, and thus the freckles become visible to the eye. Hives are common skin bumps, usually the result of an allergy or some other unknown factor that causes the blood plasma to leak out of the blood vessels and into the skin; this leakage occurs as a result of a reaction to histamine and other body chemicals. Warts are still another common type of skin lesion. They are caused by the human papilloma virus, which can be passed from person to person. This virus causes the cells in the skin to increase, with subsequent thickening of the outer covering of the skin in a specific area.

Athlete's foot is the most common fungus infection of the skin. It typically involves the web between the fourth and fifth toes, but can include any or all toes, and even the sole of the foot. The affected areas burn and itch, and the skin itself may look macerated and scaly. Good hygiene is important for prevention, as is avoidance of infected floors or showers and keeping the feet dry.

Psoriasis is a common inflammatory disease of the skin that is often chronic. There may be a loss and/or thinning of one or more of the epidermal layers, the keratin may be abnormal, white blood cells are found in the epidermis, and there are dilated blood vessels in the dermis. The skin is affected with plaques and covered with silvery scales. Most commonly affected are the knees, scalp, and elbows. The cause is unknown, but heredity may play some part.

Most experts now believe that exposure to sunburns or sun damage is responsible for most of the different skin cancers. Some skin cancers can be easily treated, and the person will have no further problems. Other skin cancers can be fatal. The latter is true for melanoma, the most malignant of all skin cancers, which affects 32,000 and kills 6,500 people in a given year. It is the eighth most common cancer, and develops when melanocyte cells, those that give color to your skin, go awry. This type of cancer is particularly deadly. Its incidence is increasing, and it appears that this may be related to increased sun exposure and depletion of the protective ozone layer in the earth's atmosphere. The number of people who develop malignant melanoma is greater among those who are white and living near the equator. Other people that are susceptible are those with a fair complexion, light hair, blue or light eyes, freckles, a family member who has had it, or a history of a blistering sunburn in the first two decades of life (which doubles the risk of melanoma later in life).

Turn the page for a look, up-close and personal, at the skin, hair, and nails and some of the more common skin conditions.

The Skin, Hair, Nails, and How Your Skin Repairs Itself

The Skin and Hair

This figure shows a magnified version of the skin and its three layers—the epidermis, dermis, and the subcutaneous layer. The hair, glands, and other components are also depicted.

Hair is a derivative of the epidermis of the skin and is made up of the protein keratin. A strand of hair consists of a shaft and a root lying within the epidermis, the hair follicle. The end of the root expands to form a hair bulb. Each hair shaft has three zones—the cuticle, cortex, and the medulla. The cuticle forms the hair surface and consists of keratin. The cortex contains many closely packed, elongated cells, which are also filled with keratin and melanin-containing cells. The central medulla consists of rounded cells containing vacuoles, keratin, and melanosomes (melanin-containing cells). The growth rate of hair varies with the area and thickness of the hair, but generally averages from 1.5 mm to 2.2 mm a week. It also varies with the cycles of growth and hair loss.

The *sebaceous glands* produce an oily substance, sebum, which keeps the skin lubricated.

When *arrector pili muscles* contract, as from cold or emotions, they elevate the epidermis and cause dimpling of the skin, or "goose bumps." These muscles also help to express secretion of the sebaceous glands, which lubricate the hair.

The eccrine *sweat glands* are dispersed in the skin and help to control the body temperature. Another type of sweat gland is the apocrine gland, which is less coiled and less widely distributed. Apocrine glands are usually found in the axilla, genital, perianal, umbilical, and external ear areas.

The *epidermis* is thin and relies on the *dermis* for nutrition. The epidermis is constantly renewed from the bottom up. New cells are formed and push their way up, moving older cells to the surface, where dead cells are sloughed off. This complete process—a cell moving up through the epidermis and being sloughed off—generally takes about 45–75 days, less in areas where the skin is thinner.

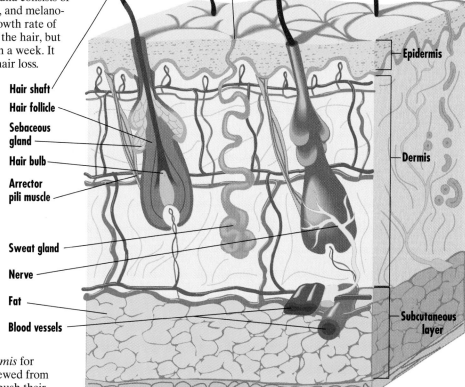

Pore

Hair shaft

Hair follicle

Sebaceous gland

Hair bulb

Arrector pili muscle

Sweat gland

Nerve

Fat

Blood vessels

Epidermis

Dermis

Subcutaneous layer

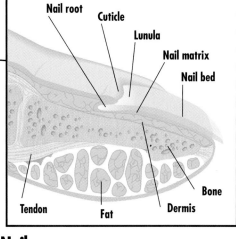

Nail root
Cuticle
Lunula
Nail matrix
Nail bed
Tendon
Fat
Dermis
Bone

Nails

Your finger and toenails consist of dead cells made up of keratin. The nails contain three major regions—the root, the body of the nail, and the free portion extending over the finger or toe tip. The nail root is inserted into an overlying cuticle. It rests on a plate of living epidermal cells, the matrix, which is joined by the dermis tissue to the finger bone. The matrix extends for some distance beneath the nail's body, where it contributes to the crescentic area, the lunula. The area beneath the nail is rich in blood vessels, and that is why the area directly beneath your visible nail looks pinkish in color. The rate of growth is most rapid in the middle finger—about 0.1 mm per day, and is slowest in the little finger.

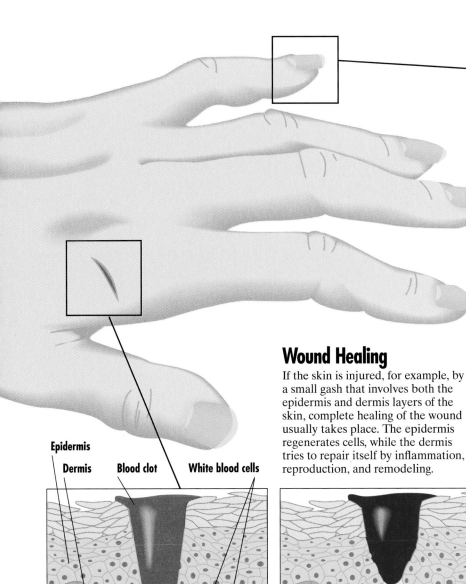

Wound Healing

If the skin is injured, for example, by a small gash that involves both the epidermis and dermis layers of the skin, complete healing of the wound usually takes place. The epidermis regenerates cells, while the dermis tries to repair itself by inflammation, reproduction, and remodeling.

Epidermis
Dermis
Blood clot
White blood cells

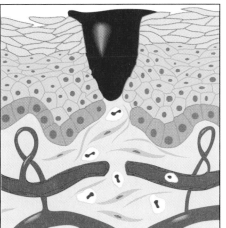

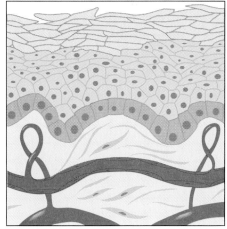

Inflammation

1 During inflammation, bleeding in the skin is stopped and is clotted off. White blood cells come to the site to kill off bacteria and to release substances to produce new tissue. Damaged tissues are also taken away.

Reproduction

2 During reproduction, cells and other substances increase, which produces more new tissue and also new blood vessels—capillaries. These substances form a matrix for the second new tissue to fill in the wound.

Remodeling

3 During remodeling, the first new tissue is replaced by scar tissue, with an abundance of collagen, which increases the strength of the wound site.

Moles and Age Spots

Skin moles are areas of the skin where the melanocytes are packed together tightly in a specific area. One of the most common benign skin moles is called seborrheic keratosis and occurs most frequently after middle age, and sometimes develops rapidly in crops. Most moles remain harmless. However, if they are exposed to excessive sunlight, trauma, or abrasion, they may become cancerous. If they become malignant, they can break through the surrounding skin, multiply in other areas of the skin, and possibly travel to other parts of the body and "seed" cancer to other areas. Everyone has skin moles, but when are they cancer? You should consult your doctor if the mole has any of the following characteristics, presented in the format of the American Cancer Society's ABCD's: If the mole is asymmetric, has irregular borders, has color variations (brown, black, white, red, or blue), or the diameter is greater than 6mm (the size of a pencil eraser).

Moles

A mole or a benign (harmless) nevus is a colored, raised area on the skin. Some moles are present from birth, and others develop later. The hallmarks of benign moles are regularity of color, an even surface, and an even border.

Age Spots

Age spots are common, harmless lesions usually found in older people. They are typically found on the back, chest, or face. They appear as plaques that have an almost "stuck-on" appearance, and are brown or sometimes very dark in color.

Skin Cancer

The American Cancer Society recommends these methods to help prevent all skin cancers: Avoid exposure to strong sun, don't use tanning booths, wear clothes to cover you and use a sunscreen lotion when out in the sun, regularly examine your skin for abnormalities, and if you see a suspicious mole, contact your doctor immediately.

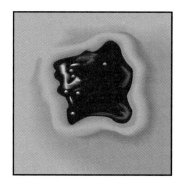

Malignant Melanoma

Malignant melanoma is the most important skin cancer because the risk of cancer spread, or metastases, and death is high, unless it is detected and treated at a very early stage. Malignant melanoma can spread to the liver, lungs, bones, eyes, or brain. It has been estimated that by the year 2000, the lifetime risk of developing malignant melanoma may be as high as 1 out of every 75 to 90 people. In men, the cancer typically affects the back and trunk, and in women, the legs, thighs, back, or upper arms.

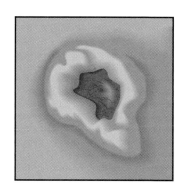

Basal Cell Cancer

Basal cell cancer is the most common of all skin cancers, with approximately 600,000 cases per year. Fortunately, this cancer has a low potential for spreading to other parts of the body, unlike malignant melanoma. The causes can be many, although it is most often associated with exposure to sunlight, fair complexion, a history of burns, ulceration, or certain scars. Basal cell cancer typically appears on the face or neck and frequently has a pearly border. It slowly increases in size and may become ulcerated.

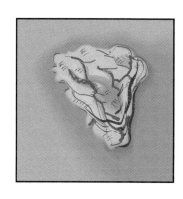

Squamous Cell Cancer

About 20% of all skin cancers that are not melanomas are squamous cell cancer, and they do have a potential to grow rapidly and to spread to other body parts. Squamous cell cancers also typically occur on the face, especially the lower lip and ears, and often exhibit ulcerations and crusting. The causes include sun exposure, burns, certain viruses, and immunosuppression.

CHAPTER

6

The Bones

YOUR BONES ARE important because of the structure they provide, and also because of what they contain. They are the framework for the rest of your body. Together they give your body shape, and—moved by your muscles—allow you to run after that tennis ball, take a romantic walk in the woods, or type a letter to your mom. They protect delicate body parts. They also store bone marrow to produce blood cells and contain minerals like calcium and phosphorus that can be released into the blood. It is because of these minerals that bones can survive for many years, even after death.

You have 206 bones in your body, many of which are held in place by strong bands of tissues called ligaments. Both men and women have the same number of bones and the same basic bone structure. However, men usually have broader shoulder bones and a longer rib cage, while women have wider pelvic bones.

The framework formed by your bones is your skeleton, which is divided into three parts. The axial skeleton consists primarily of your vertebral column, skull, ribs, and sternum. The appendicular skeleton consists of the bones of your arms and legs. The ossicle bones in your ears form the third skeletal part (see Chapter 3).

Your bones, as the total skeleton, are able to move because of the subsequent joints that are formed by two bones meeting. Your body parts can have different types of movements depending on the joints involved. For instance, your shoulders and hips have the most freedom of movement because their joints form a ball and socket. Your elbow joint is considered a hinge joint, having a considerable degree of movement but only able to move in one plane or direction. The simplest type of joint movement, usually limited by ligaments or bony processes (protrusions of bone), is a gliding motion typically found in vertebrae and some hand and foot bones.

Now, let us learn more about some of the specific bony areas in your body. The largest bone in your body is the femur or thigh bone, and the smallest bone is the stapes of the middle ear. The skull, or bony part of the head, is divided into two parts—the cranium, which has eight bones and protects the brain, and the skeleton of the face, which consists of 14 facial bones. The vertebral column or backbone comprises the 33 bones called vertebrae. They help support your body, are the basis of your posture, and encase and protect your delicate spinal cord. Individual nerves come out from the spinal cord, go through openings in the vertebrae, and supply the organs and tissues of your body with valuable information. For example, nerve impulses can be sent to the muscles in your hand to pick up a glass of water, to your legs to run to catch a bus, or to your ribs and diaphragm to assist in breathing. There are seven vertebrae in your neck or cervical area, 12 in your upper-middle back or thoracic area, five in the lower back or lumbar area, five in the pelvic or sacrum area, and four in the tailbone or coccyx area.

Your upper limbs consist of the bones of the arms and hands and the collarbone and shoulder blade, which make up the shoulder. Your lower limbs consist of the hip bones and the bones of the legs and feet. The upper limb bones are designed to give you dexterity, while the pelvis and the leg bones are important for walking and bearing your body's weight.

There are four main types of bones. Short bones are found in the wrist and ankle, flat bones are found in the skull and ribs, long bones are found in the arms and legs, and irregular bones are found in the spine and skull. Most of these bones consist of three basic layers. The first is a thin membrane, called the periosteum, which has nerves and blood vessels. The next is a hard, compact layer called cortical bone. The innermost layer is called cancellous bone. It is spongy, and its inner cavity contains bone marrow, which produces your red and white blood cells.

New bone tissue is constantly being formed and also broken down within your bones. This is achieved by the two basic types of bone cells. Osteoblast cells help to build bone tissue by adding calcium, while osteoclasts break down bone tissue and release calcium into the blood. Your hormones influence this build-up or breakdown of the bones. The female hormone estrogen, the male hormone testosterone, growth hormone from the pituitary gland, and calcitonin from the thyroid glands stimulate the formation of new bone, while the parathyroid hormone from the parathyroid glands increases bone breakdown. Up until the age of about 35, your bones are constantly being remodeled, and the amount of new bone being created is greater than the amount of old bone being broken down. After age 35, your bones have reached their maximum strength and density, and more bone material is lost than is created.

Osteoporosis is a disorder in which calcium and other elements of bone are abnormally depleted, the bone mass is reduced beyond the usual range, and the bone is more brittle. There are two primary forms of osteoporosis. The most common occurs primarily in women after menopause, and the other occurs in men and women over the age of 70. In many cases osteoporosis is a process that begins early in adult life and progressively worsens; this is especially true in women after menopause. Although osteoporosis can affect men, it primarily affects women over the age of 55. It also targets Caucasians more than African Americans or Asians. The primary reason that women develop osteoporosis after menopause is that the female hormone estrogen, essential for women's bone development, is dramatically decreased after this time. Men have very little estrogen, and the male hormone testosterone, important for their bone development, does not usually decrease precipitously as they age. However, men can develop osteoporosis from conditions related to old age, alcoholism, or other factors. Men or women of any age can also develop osteoporosis from such causes as certain drugs, illnesses, or deficiencies.

People with osteoporosis are predisposed to breaking their bones—most likely, the hips, wrists, arm or leg bones, or the spine. Often, especially if the disease is mild or in an early stage, there may be no symptoms. If the disease is more advanced, the person may become shorter because of collapse of the vertebrae and may suffer from vague backaches, slumped posture, and fractures of the bones. Osteoporosis can usually be diagnosed by the use of x-rays and other tests to measure bone density. Though a tendency to develop osteoporosis may be inherited, in many cases it can be prevented by such measures as adequate calcium intake during one's life, regular exercise, and avoidance of cigarettes and rapid weight loss.

Other common conditions related to the bones are arthritis, scoliosis, neck and backaches, and vertebral bone injuries. There are several kinds of arthritis, of which the two most common are osteoarthritis and rheumatoid arthritis. Osteoarthritis, the most common degenerative joint disease, occurs primarily in older people and most frequently affects the fingers, hips, knees, and spine. It produces pain and stiffness of these joints. Rheumatoid arthritis is an autoimmune disease in which the person's immune system attacks the joints, resulting in inflammation, degeneration, and/or derangement. It can occur in people of any age and typically affects the hands, wrists, knees, and feet. The symptoms typically include pain, stiffness, or limitation of movement of the affected parts. Both types of arthritis usually afflict women two to three times more than men.

Scoliosis is an abnormal curvature of the spine, usually recognized in childhood. It is believed that it may result from abnormalities in the muscles of the spine or in the position of the bones. Most people with scoliosis do not have any symptoms and do not require treatment. However, severe cases can produce back pain, shortness of breath, and occasionally pressure on the lungs or heart.

Neck aches can arise from many situations—serious and not so serious. Usually, they are the result of sleeping in an awkward position or a temporary spasm. They can also result from whiplash or other injuries, arthritis, or a herniated disc. A herniated disc occurs when a disc between your vertebrae ruptures, and the ruptured part slips out of place, causing pressure on a nerve and causing pain. Back pain is most commonly found in the lower region of the spine and is typically due to an injury or stretch of a muscle, ligament, or nerve. It can also be the result of arthritis or herniated discs.

Fractured vertebrae of the spine—from car accidents, sports injuries, or other forms of trauma—are very serious. If this type of injury is suspected, the person should not be moved except by trained medical people. Symptoms can include severe pain, inability to feel sensations, or inability to move.

As you turn the page, you will see just how your bones work—inside and out.

The Skeleton

The Skeleton from the Front
Some of the major bones, as seen from the front of the body, are shown.

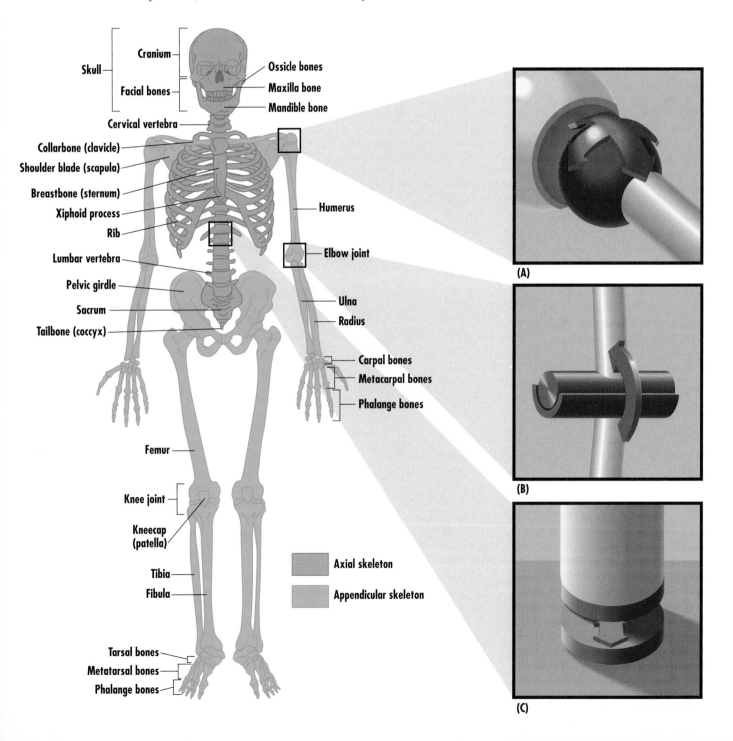

Skull
Cranium
Facial bones
Ossicle bones
Maxilla bone
Mandible bone
Cervical vertebra
Collarbone (clavicle)
Shoulder blade (scapula)
Breastbone (sternum)
Xiphoid process
Rib
Humerus
Elbow joint
Lumbar vertebra
Pelvic girdle
Sacrum
Tailbone (coccyx)
Ulna
Radius
Carpal bones
Metacarpal bones
Phalange bones
Femur
Knee joint
Kneecap (patella)
Axial skeleton
Appendicular skeleton
Tibia
Fibula
Tarsal bones
Metatarsal bones
Phalange bones

(A)

(B)

(C)

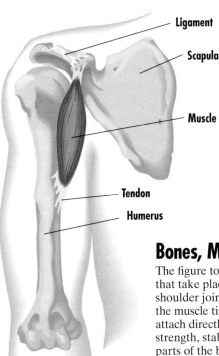

Ligament

Scapula

Muscle

Tendon

Humerus

Bones, Muscles, and Tendons

The figure to the left depicts the connections that take place in your body, in this case the shoulder joint. As muscles approach bones, the muscle tissue changes into tendons, which attach directly to the bone and provide extra strength, stability, and movability. In some parts of the body, movability is not as important as stability and security, and so ligaments connect one bone to another.

The Joints

The figures to the left show mechanical depictions of three different types of joints.

Figure A depicts the shoulder, a ball-and-socket joint and the most moveable type of joint. The shoulder is formed by the head of the humerus bone fitting into the cup-like joint cavity, thus the name ball-and-socket joint. The shoulder allows you to move your arm toward or away from your body, rotate your arm, or move it in circular motions.

Figure B illustrates the elbow, a hinge type of joint. The elbow allows you to extend, flex, and move your forearm and hand so that your palm faces up or down.

Figure C shows the spine's bones or vertebrae, which are a gliding type of joint. The vertebrae glide on top of each other, making flexion, extension, and rotation possible.

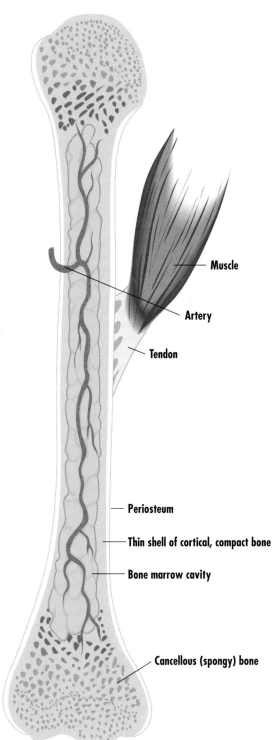

Muscle

Artery

Tendon

Periosteum

Thin shell of cortical, compact bone

Bone marrow cavity

Cancellous (spongy) bone

The Structure of Bone

Bone has three layers—the periosteum, or membrane for nerves and blood vessels; the compact or cortical bone; and the innermost cancellous bone, with the bone marrow cavity where blood cells are formed.

Osteoporosis

Causes of Osteoporosis

The most common cause of osteoporosis is a decrease in the amount of estrogen hormone as a result of menopause in women. Other causes are:

Gland abnormalities (such as estrogen deficiency from menopause, testosterone deficiency, or an increase in thyroid or parathyroid hormones)

Confinement to bed

Poor diet (deficiency in calcium, protein, etc.)

Medicines (such as steroids) or alcohol

Prolonged use of casts (osteoporosis occurs at the site)

Paralysis

Genetic diseases

Space travel

Bone tumors

Chronic disease

The Vertebrae and Osteoporosis

The bones of osteoporotic people are more fragile because less bone mass is present. The vertebrae of the spine lose their height and collapse, sometimes leading to compression of surrounding nerves, fractures of bones, and severe pain.

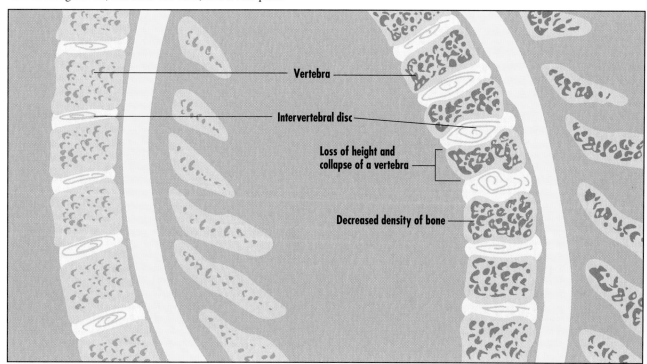

Vertebra

Intervertebral disc

Loss of height and collapse of a vertebra

Decreased density of bone

Normal spine

Osteoporotic spine

Osteoporotic Bone

These illustrations show how osteoporotic bone (right) has a "moth-eaten," more barren appearance compared to normal bone (left).

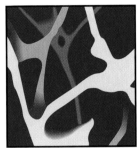

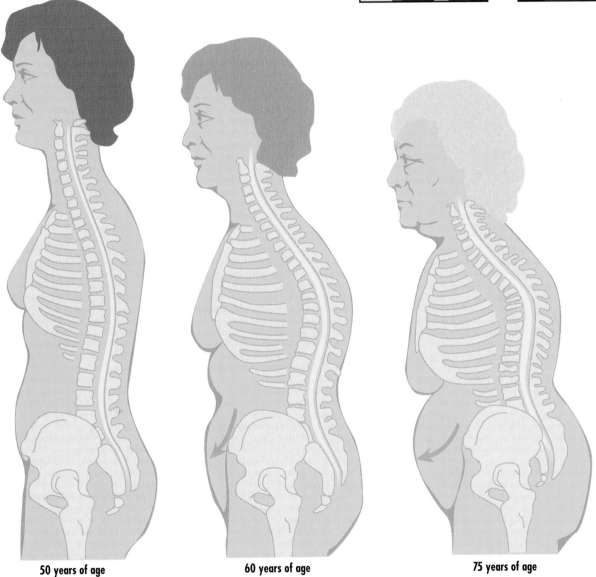

50 years of age   **60 years of age**   **75 years of age**

Osteoporosis and Body Posture

As osteoporosis continues to develop, not only do the vertebrae lose their height, but the person actually becomes shorter. There is a progressive slumping of the shoulders and back, which is called a dowager's hump. If this slumping continues, the lower ribs eventually lie on the hip bone, and the resulting pressure changes cause a ballooning of the abdomen.

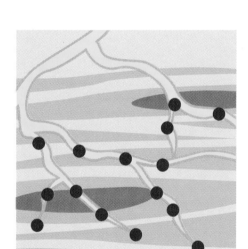

CHAPTER
7

The Muscles

THE MUSCLES IN your body are crucial to various types of movement—voluntary and involuntary. Without muscles, Michael Jordan could not have soared on the basketball court, Fred Astaire could not have danced the waltz, Janet Jackson or Frank Sinatra could not sing, and you could not swallow a delicious meal or laugh at a joke. Muscles help you act and react in your world.

There are three different muscle types—skeletal, smooth, and heart. Skeletal muscles make up about 40% of your body and are generally under your conscious control. They primarily help to regulate movements of the head, neck, back, arms, and legs, and in doing so, they control posture and help you move about. Heart and smooth muscles make up about 10% of your body weight. The heart or cardiac muscles are found only in the heart and have a slower contraction than the other muscles. The heart muscles are controlled by your heart's own pacemaker (see Chapter 10) and are not under your willful control. Smooth muscles also are not under your voluntary control. They line organs and other tissues such as the stomach, intestines, blood vessels, bronchi, gallbladder, urinary bladder, pupil of the eye, and anus. Obviously, smooth muscles are involved in varied processes—digestion, control of blood pressure, and control of urine release, for example.

Though similar in basic ways, the three muscle types do perform differently. Skeletal muscles always have two points of attachment to bones—one called the origin and one called the insertion. This is important for fixation and movement of the muscles. Your muscles are attached to these bones via dense, white, glistening bands called tendons. Nerves are usually embedded within the muscles, and they transmit important information to the muscles about what the body should do. Muscles move thanks to these nerve messages, which can be triggered by many different stimuli. A less intense nerve message produces a weaker muscle response of short duration, and a more intense nerve message initiates a longer and stronger muscle response.

The movement of your skeletal muscles takes place in the following manner. A nerve carries messages from your brain to the muscle fibers, releasing a substance called acetylcholine, which acts on muscle fibers to create what is called an action potential. An action potential is similar to an electrical stimulus and also sends a message—in this case, causing large amounts of sodium ions and then later calcium ions to be released into the muscles. These ions, in turn, produce attractive forces of the actin and myosin within the muscle fiber. This attraction causes the actin and myosin to slide together, the muscle contracts, and the relevant arm or leg is able to move. Another chemical is later released to permit the actin and myosin to separate, the contraction stops, and the muscle relaxes. When death occurs, all the skeletal muscles of the body contract and become rigid, which is called rigor mortis. This is the result of the body's failure to release the chemical that normally separates the actin and myosin.

The skeletal muscles in your body are constantly remodeled in response to use, disuse, and other factors. When body builders build up specific muscle groups, they are enlarging individual muscle fibers, as well as the total mass of that muscle. This is called muscle hypertrophy. If a muscle is not used over a long period of time, the number of myofibrils (slender muscle threads within a muscle fiber) and the proteins within them decreases, the muscles become weaker and smaller, and then muscle atrophy occurs. If a muscle loses its nerve supply, as from a stroke or injury, it can no longer receive contractile messages, and atrophy can begin immediately. However, if the nerve supply grows back within three months, function can return. After that amount of time, there is less return of function, with generally no return of function after one to two years.

Like skeletal muscles, smooth muscles contain actin and myosin filaments, though smooth muscles differ from skeletal muscles in that smooth muscles have fewer myosin filaments, a slower onset to contraction, as well as a stronger and more prolonged contraction. It is because of these different properties that your stomach muscles can contract to mix, digest, and propel the food to the small intestines, and your blood vessels can be normal or dilated one minute, but very quickly constrict when you leave your warm house and go outside into the cold without a hat or coat on. Smooth muscles also rely on another component, dense bodies, for contraction. The dense bodies are attached to actin filaments and to other proteins that link one dense body to another. It is because of these attachments that the muscle contraction can be spread from one part to the entire muscle or organ. Still another difference is that smooth muscles can be made to contract not only from the action of the nerves but also from other stimuli such as with hormones, decreased body temperature, or changes in the levels of carbon dioxide or other body chemicals. This is important because it allows for rapid changes in the size and activities of the tissues and organs that are made up of smooth muscles.

Heart muscles look and act more like skeletal muscles, and they have similar actin and myosin filaments. However, heart muscles are more interconnected than those of either skeletal or smooth muscles, and because of this, heart muscle contractions can last a long time.

Disorders related to muscles can be divided into those resulting from injury (the most common) and those resulting from disease. A strained or pulled skeletal muscle (a shoulder muscle, for example) could result from stretching too long or too high when painting your house, a strained calf muscle could result from running a marathon when you're not properly conditioned, or a strained back muscle could occur from lifting a package that is too heavy. Muscle strains may produce ruptures of some fibers within the muscle tissue, which can initiate pain and stiffness. However, this commonly improves itself in a few days or weeks (depending on the severity), as the tissue repairs itself and the swelling and inflammation subsides.

A sprain does not directly involve a muscle, as some people think, but instead is due to the stretching or tearing of ligaments that hold joints together. This can cause swelling, pain, and tenderness in the vicinity, but it slowly resolves as the ligament strands grow back together, and the swelling and inflammation again dissipate. See Chapter 9 for information and an illustration of ankle sprains.

Perhaps the most well-known muscle disease is the Duchenne's type of muscular dystrophy. It is hereditary and typically becomes apparent in boys between the ages of three and six. This type is rapidly progressive, whereas other types may be less severe and progress more slowly. It occurs in about one out of every 3,500 boys. The disease initially produces enlargement of some muscles (especially the calves), but then degeneration occurs producing a decrease in the number of muscle fibers. This reduction of muscle fibers makes climbing stairs, walking, and other movements difficult or impossible. Those muscle fibers that remain become haphazardly organized, with some fibers being smaller and some larger. In the later stages of the disease, only a few scattered muscle fibers remain in a pool of fat cells.

Muscle pain or weakness are common symptoms of many diseases. Types of flu and other infections are the most frequent cause of muscle pain and weakness. However, these symptoms can also occur from the effects of rheumatoid arthritis, thyroid disease, alcohol or heroin use, or metabolic problems caused by changes in the chemicals of your body (for example, calcium, magnesium, sodium, or potassium). In one way or another, these conditions can temporarily or permanently damage the muscle fibers, resulting in abnormal contractions, weakness, or pain.

In cases where no injury has occurred, people who have muscle pain or weakness may be asked by their doctor to have some further tests. These can include blood enzyme tests, to indicate if muscle damage has allowed certain enzymes to leak; a CT (computerized tomography) scan; magnetic resonance imaging; muscle biopsy; or a test called an EMG, which measures the electrical activity of muscles.

Now, please let your finger and arm muscles do the walking and turn the page so that you can see and learn more about your muscles.

The Anatomy of Muscles

Major Muscles of the Front of the Body

Some of the major muscles of the front of the body are shown, with their primary functions.

Sternocleidomastoid—Bends the head to the same side, rotates the head, and raises the chin to the opposite side.

Serratus anterior—Raises the shoulder and is useful in pushing.

Deltoid—Flexes and rotates the arm.

Biceps—Flexes arm and forearm and turns the hand.

Brachioradialis—Flexes the forearm.

Flexor carpi radialis—Flexes and turns the hand.

Pectoralis major—Flexes and rotates the arm and is useful in climbing.

Rectus abdominis—Compresses the abdomen and flexes the spine.

External abdominal oblique—Compresses the abdomen and assists in urination, defecation, vomiting, the birth process, and in letting out a deep breath. It also flexes and rotates the spine, and depresses the ribs.

Tibialis anterior—Flexes and inverts the foot.

Rectus femoris—Flexes the thigh and extends the leg.

Vastus lateralis—Extends the leg.

Sartorius—The longest muscle in the body, flexes and rotates the thigh and leg.

Vastus medialis—Extends the leg.

Gastrocnemius—Flexes the leg and foot.

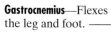

Tendons leading to origin

Belly of muscle

All muscles have two points of attachment—a point on a bone from which they originate (the origin), and another point on a bone to which they are inserted (the insertion). For stability and strength purposes, both the origin and insertion areas are generally tendons, which are fibrous cords by which a muscle is attached to bone. The belly of the muscle is the main bulk of the muscle itself.

Figure (A) shows a depiction of the biceps muscle of the arm and figure (B) illustrates the gastrocnemius muscle of the leg, both showing the areas of origin and insertion.

Tendon leading to insertion

(A)

Major Muscles of the Back of the Body

Here are some of the major muscles of the back of the body and their primary functions.

Infraspinatus—Rotates the arm.

Deltoid—Flexes and rotates the arm.

Extensor carpi radialis longus—Extends and turns the hand.

Extensor carpi ulnaris—Extends and turns the hand.

Flexor carpi ulnaris—Flexes and turns the hand.

Trapezius—Raises, turns, and lowers the shoulder and turns the face to the same or the opposite side.

Teres major—Extends and rotates the arm.

Triceps—Extends and rotates the arm and extends the forearm.

Latissimus dorsi—Extends and rotates the arm, draws the shoulder down and back, and helps in climbing.

External abdominal oblique

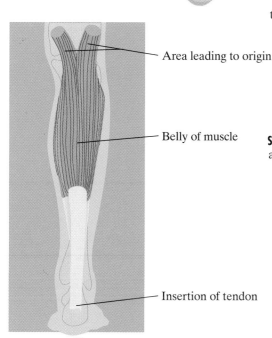

Area leading to origin

Belly of muscle

Insertion of tendon

(B)

Semitendinosus—Extends the thigh and flexes the leg. After it is flexed, can rotate it.

Gluteus maximus—Extends and rotates the thigh and braces the knees.

Biceps femoris—Extends the thigh and rotates it, and flexes the leg and rotates it.

Semimembranosus—Flexes the leg, and after it is flexed, may rotate it. Also extends the thigh.

Gastrocnemius—Forms the bulk of the calf of the leg and flexes the leg and foot.

Soleus—Flexes the foot.

Gluteus medius—Turns, rotates, and can also flex or extend the thigh.

How Muscles Function

All muscles are made up of fibers, thinner than a hair, that can contract and relax—the basic work of movement. Within each muscle fiber are hundreds to thousands of smaller muscle threads called myofibrils, and each myofibril has about 1,500 thick myosin filaments and about 3,000 thinner actin filaments. These muscle components (the actin and myosin) interact to cause muscle contraction and relaxation.

Skeletal muscles are crucial for movements and are found in areas such as the arms, legs, back, and chest and abdominal walls. They are usually under your direct control. Smooth muscles are found in blood vessels and organs and are not consciously controlled by you.

Voluntary (Skeletal) Muscle Movement

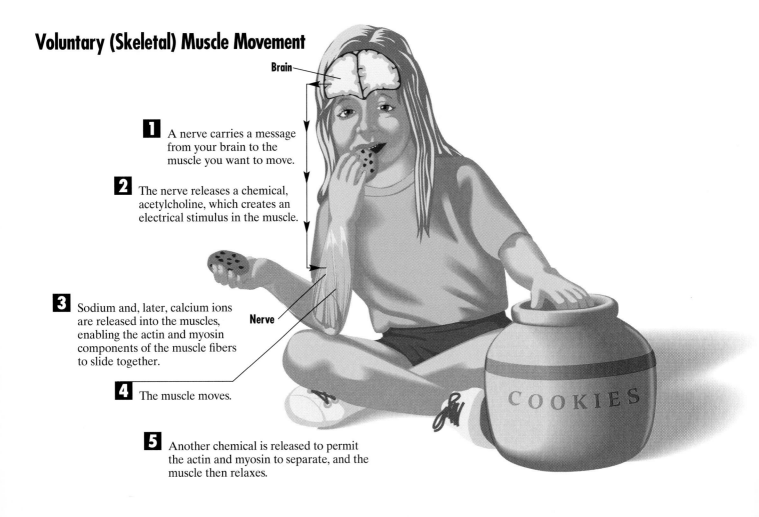

Brain

1 A nerve carries a message from your brain to the muscle you want to move.

2 The nerve releases a chemical, acetylcholine, which creates an electrical stimulus in the muscle.

3 Sodium and, later, calcium ions are released into the muscles, enabling the actin and myosin components of the muscle fibers to slide together.

Nerve

4 The muscle moves.

5 Another chemical is released to permit the actin and myosin to separate, and the muscle then relaxes.

COOKIES

Involuntary (Smooth) Muscle Movement

Contraction is much slower in smooth muscles than in skeletal muscles, but smooth muscles can produce prolonged contractions for hours or even days, whereas the contraction in skeletal muscles is much more brief. Smooth muscles do not have the same arrangement of actin or myosin filaments that skeletal muscles do. Instead, the actin filaments of smooth muscles are attached to structures called dense bodies. Some of these dense bodies are linked to other dense bodies by protein filaments that become bridges. It is through these bridge bonds that contraction takes place.

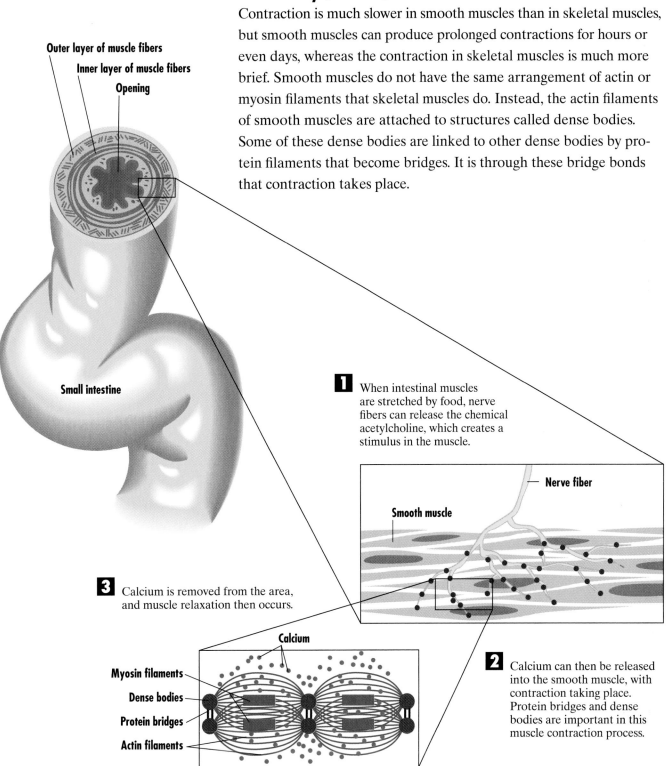

Outer layer of muscle fibers
Inner layer of muscle fibers
Opening

Small intestine

1 When intestinal muscles are stretched by food, nerve fibers can release the chemical acetylcholine, which creates a stimulus in the muscle.

Nerve fiber

Smooth muscle

3 Calcium is removed from the area, and muscle relaxation then occurs.

Calcium

Myosin filaments
Dense bodies
Protein bridges
Actin filaments

2 Calcium can then be released into the smooth muscle, with contraction taking place. Protein bridges and dense bodies are important in this muscle contraction process.

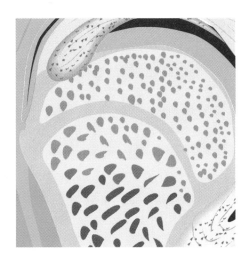

CHAPTER

8

The Arms, Forearms, and Hands

YOUR UPPER EXTREMITY consists of your arm, forearm, and hand. The major joints involved are the shoulders, elbows, and wrists. Considering all the components of your upper extremity, your hand is one of the most complex, flexible, and also highly sensitive structures—particularly to pain, temperature, and sensation. It is through your hands, and then ultimately your arms, that you learn more about your world. For example, as a child you come to realize why it's not a good idea to touch a hot stove, you determine the pleasurable sensations of feeling a soft, thick piece of velvet, and you learn how to grasp and throw a baseball. As you grow older, your hands can bring you into the musical world of playing the piano, your hands can quickly type a letter into your computer, or they can give a back massage after a long day.

The largest bone of the entire upper extremity is the humerus, and it is the only bone of your upper arm. There are two bones in your forearm—the radius, which runs in a line above the thumb, and the ulna, which runs in line with the little finger. Each hand consists of eight carpal bones (some of which help to form the wrist joint), five metacarpal bones, and 14 phalange bones (two for the thumb and three for each finger).

The three joints of your upper extremities—the shoulder, elbow, and wrist—are crucial to the movements you need to function easily. The shoulder joint is the junction of your arm and body and is held in place by strong ligaments. It is made up of the clavicle, scapula, and humerus bones, and is attached to the chest by multiple muscles and the sternoclavicular joint. There are three basic subjoints within the shoulder itself—the sternoclavicular (formed by the sternum and clavicle bones), the glenohumeral (formed by the head of the humerus bone and the glenoid cavity of the scapula), and the acromioclavicular (formed by the sternum and clavicle bones). The shoulder is a ball and socket joint, and this type of joint is the most moveable of any in your body. Altogether, your shoulder movements allow you to move your arms toward or away from your body and rotate them inward or outward and in circular motions. The glenohumeral joint is the most moveable part of the shoulder and has a loose capsule. The other areas of the shoulder and its connections are less moveable and have tight ligaments to hold them together. Most of the strength of the shoulder results from the muscle tendons above, behind, and in front of it.

The elbow is a hinge type of joint between the humerus bone and the bones of the lower arm, the radius and the ulna. When the biceps muscle contracts, the elbow, and thus the arm, bends. When the triceps muscle contracts, the elbow straightens. Four types of motion can be achieved by your elbow—extension, flexion, and movement of your forearm and hand so that your palm faces up or down.

The wrist and hand are the most active parts of your upper extremity. The palm side of your hand comes in the most contact with objects, and because of that, its surface has some special features. The skin of the palm side of your hand is thicker and has a greater number of blood vessels than the back of your hand; it contains no hairs, but it does have numerous sweat glands. The large number of hand muscles and the rich supply of nerves to the hands allows them to produce the most intricate movements of your body, which no other animal can achieve. Of all creatures, only you (and other humans) are able to clench, grasp, squeeze, pinch, flex, extend, oppose, and move your fingers in and out.

The wrist is a joint between the radius and ulna bones, and the hand or carpal bones. The wrist is a condyloid type of joint—that is, one oval-shaped bone fits into the elliptical cavity of another bone. The possible movements of your wrist include flexion, extension, and movement toward or away from your body. Within the inner part of your wrist is a cavelike space called the carpal tunnel, which is bounded by four of the carpal bones (some of which help to form the wrist joint). The flexor retinaculum, a horizontal band of strong fibrous tissue, runs between these four bony areas and forms a sheath, or tunnel. The carpal tunnel transports the median nerve and finger muscle tendons from the forearm to the hand, and it is important not only because of these tissues that lie within it, but also because of the problems that can result if the area becomes swollen or inflamed.

Carpal tunnel syndrome is a wrist disorder resulting from the swelling of this flexor retinaculum. This swelling causes compression of the median nerve, resulting in pain, tingling, numbness, and/or weakness in the wrist, hand, and fingers. The problem is more common in women, and this in part may be due to hormonal changes, as carpal tunnel syndrome often occurs during pregnancy, menopause, or before the menstrual period, when additional fluids may collect in the body and its ligaments. This disorder can also be caused by, and worsens with, the excessive use of the wrist, especially in repetitive activities, such as sewing or computer work. Over time, these repetitive actions can produce injury and/or swelling of the flexor retinaculum and nerve, and may thereby cause the syndrome to occur. Carpal tunnel syndrome may also occur as a result of fluid collection after injury, infections like tuberculosis, vitamin B6 deficiency, or arthritis. Often the symptoms can affect both hands. Symptoms may travel into the palm surface of the thumb, index finger, middle finger, the inner half of the ring finger, and may even spread into the arms. These symptoms usually occur or worsen at night.

In addition to the wrist disorder carpal tunnel syndrome, some other common upper extremity joint problems that involve the shoulder and elbow are dislocations of the shoulder, frozen shoulder, and tennis elbow. Shoulder dislocations or displacements may affect any of the basic subjoints within the shoulder area, and they usually force the corresponding ligaments and/or bones out of proper position. Mild dislocations usually cause only a sprain (or stretching) of the ligaments holding the joint together, and the position of the joint does not move very much out of proper alignment. A more moderate dislocation causes an actual rupture of the ligaments, with the two bony parts moving more out of alignment. In a severe dislocation, all of the related ligaments can be ruptured, and the two bony parts can be seriously displaced.

Shoulder dislocations, especially of the glenohumeral and acromioclavicular joints, occur frequently in athletic pursuits. Glenohumeral dislocations usually occur because of an extension, rotation, or pulling force exerted on the arm, while acromioclavicular dislocations usually result from a fall on the shoulder.

Participating in sports may cause a variety of other shoulder injuries as well. Both baseball and football require complex movements that can maximally stretch the shoulder joint capsule and muscles, sometimes causing tears of the muscles or other ligaments. Golf, which is usually considered a nontraumatic sport, may also result in injuries to the acromioclavicular shoulder joint. On the other hand, the underhand motion of a softball throw or bowling are not as forceful, and thus, injuries resulting from these activities are less common.

Frozen shoulder is another shoulder abnormality. This may result from an injury or a strain and may follow bursitis, tendinitis, or immobilization. Other possible causes include diabetes, heart attack, and some drugs. It produces pain and stiffness and is caused by the inflammation of the joint capsule. Frozen shoulder occurs more frequently in women, especially after the age of 50.

Tennis elbow is an injury that frequently affects middle-aged men. It is a result of a repetitive strain to the area around the ligaments of the outside edge of the elbow. It usually produces a gradual type of elbow pain and is generally found only in one elbow—typically the right. Not only tennis players, but also carpenters, plumbers, gardeners, dentists, and politicians (from excessive handshaking) are among those who often suffer from this injury.

People can be physically challenged because of impairments of any part of the upper extremity—loss of an arm from an injury, severe arthritis limiting movements of the hands, or a stroke affecting the arm(s), for example. Depending on the type and severity of the problems, people can adjust and compensate in various ways. Special equipment like a long-handled shoehorn, a gripping instrument to more easily grasp and open jars, or even a voice-activated computer are available. Additionally, some people learn to develop greater dexterity in their feet or mouth so that they can write or even paint pictures using their toes or lips. This dexterity may occur as the result of increased strength of the muscles in these areas, and/or more or better nerve transmission.

Please turn the page to learn more about the insides of your upper extremities—and some of the most common problems experienced with them.

The Upper Extremity—Shoulder, Elbow, Wrist, and Hand

The Upper Extremity

These figures show some of the important muscles, arteries, and nerves of the upper extremity.

Deltoid muscle—Important for flexing and rotating the arm inward, and extending and moving the arm away from the body.

Biceps muscle—Flexes the arm and forearm, and turns the palm of the hand to face upward.

Supinator muscle—Turns the palm of the hand to face upward.

Brachioradialis muscle
Flexes the forearm.

Flexor carpi radialis muscle
Flexes the hand, and moves it away from the body.

Palmaris longus muscle—
Flexes the wrist and hand area.

Coracobrachialis muscle—Flexes and moves the arm inward toward your body.

Triceps muscle—Extends the arm and forearm, and can move the arm inward toward the body.

Brachialis muscle—Flexes the forearm.

Bicipital aponeurosis—Tissue formation where the two portions of the biceps muscle come together.

Pronator teres muscle—Turns the palm of the hand so that it faces downward.

Flexor carpi ulnaris muscle tendon—Flexes the hand, and moves it toward the body.

Flexor digitorum superficialis muscle—Flexes the first and second bones of each finger and the hand.

Flexor retinaculum—A strong band of tissue that is important in the carpal tunnel syndrome.

Flexor pollicis longus muscle—Flexes the first and second bone of the thumb, and flexes the first metacarpal bone.

Musculocutaneous nerve

Median nerve

Ulnar nerve

Brachial artery

Radial nerve

Ulnar artery

Radial artery

Superficial arterial arch

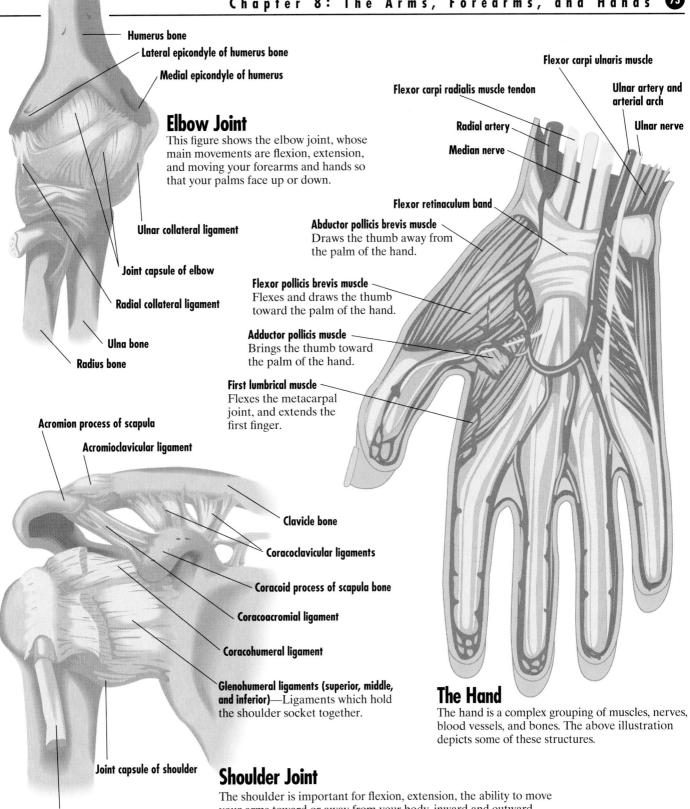

Humerus bone

Lateral epicondyle of humerus bone

Medial epicondyle of humerus

Flexor carpi ulnaris muscle

Flexor carpi radialis muscle tendon

Ulnar artery and arterial arch

Radial artery

Ulnar nerve

Median nerve

Elbow Joint

This figure shows the elbow joint, whose main movements are flexion, extension, and moving your forearms and hands so that your palms face up or down.

Ulnar collateral ligament

Joint capsule of elbow

Radial collateral ligament

Ulna bone

Radius bone

Flexor retinaculum band

Abductor pollicis brevis muscle
Draws the thumb away from the palm of the hand.

Flexor pollicis brevis muscle
Flexes and draws the thumb toward the palm of the hand.

Adductor pollicis muscle
Brings the thumb toward the palm of the hand.

First lumbrical muscle
Flexes the metacarpal joint, and extends the first finger.

Acromion process of scapula

Acromioclavicular ligament

Clavicle bone

Coracoclavicular ligaments

Coracoid process of scapula bone

Coracoacromial ligament

Coracohumeral ligament

Glenohumeral ligaments (superior, middle, and inferior)—Ligaments which hold the shoulder socket together.

The Hand

The hand is a complex grouping of muscles, nerves, blood vessels, and bones. The above illustration depicts some of these structures.

Joint capsule of shoulder

Shoulder Joint

The shoulder is important for flexion, extension, the ability to move your arms toward or away from your body, inward and outward rotation, and circular motion.

This figure shows the ligaments of the shoulder. The ligament names indicate where the ligaments come from and go to. For instance, the acromioclavicular ligament means that the ligament goes from the acromion process of the scapula bone to the clavicle bone.

Biceps muscle tendon

Disorders of the Arms and Hands

Shoulder Dislocation

Shoulder dislocation is a common injury, especially in young men and women who participate in sports. This injury frequently occurs as a result of a fall on the edge of a shoulder or a strong force on the arm.

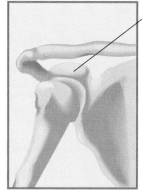

Subcoracoid dislocation

This illustration shows a type of common shoulder dislocation, called a glenohumeral dislocation, which involves the shoulder capsule and socket. It depicts a sub-type of this dislocation, which is also common, and is called a subcoracoid dislocation.

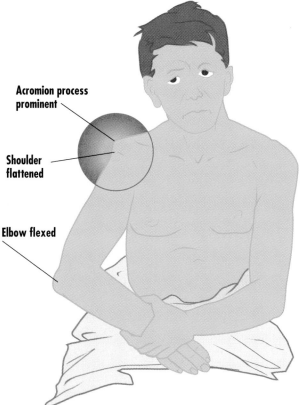

Acromion process prominent

Shoulder flattened

Elbow flexed

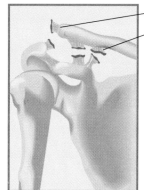

Acromioclavicular ligament torn

Coracoclavicular ligaments torn

This figure shows a shoulder dislocation involving an injury to two different types of ligaments. There can be different degrees of severity of all dislocations. For example, a ligament can be stretched or torn, or there may be a more severe injury, as shown here, to the left.

This illustration shows what a shoulder dislocation may look like. The shoulder may be flattened, the humerus arm bone can be prominent, the forearm can be rotated, and the person may use the other hand to support the elbow.

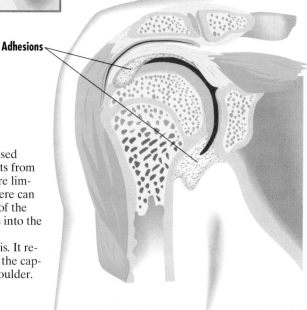

Adhesions

Frozen Shoulder

A person with a frozen shoulder, or capsulitis, has a decreased range of motion of the shoulder. Capsulitis frequently results from injuries, strains, or being immobilized. The joint motions are limited and painful, particularly on rotation. From the rear, there can be a decrease of the muscle mass on the scapula bone and of the deltoid muscles. In the more severe forms, the pain extends into the neck or down the arm.

Here is an inside view of a shoulder affected by capsulitis. It reveals that there are thickenings and inflammation between the capsule, and these adhesions in turn limit the motion of the shoulder.

Carpal Tunnel Syndrome

Carpal tunnel syndrome results from a swelling of the band of tissue lying across the wrist, called the flexor retinaculum. This swelling can then produce compression of the median nerve, which is underneath the flexor retinaculum. At the wrist, the median nerve and muscle tendons pass through a so-called tunnel whose walls are bounded by the carpal bones, and are enclosed by this flexor retinaculum. Carpal tunnel syndrome may involve both wrists and can cause abnormal sensations and weakness in the thumb, index, middle, and the first half of the ring finger.

The symptoms of this syndrome can be reproduced by doctors when they tap on the affected area, or by juxtaposing the hands back to back.

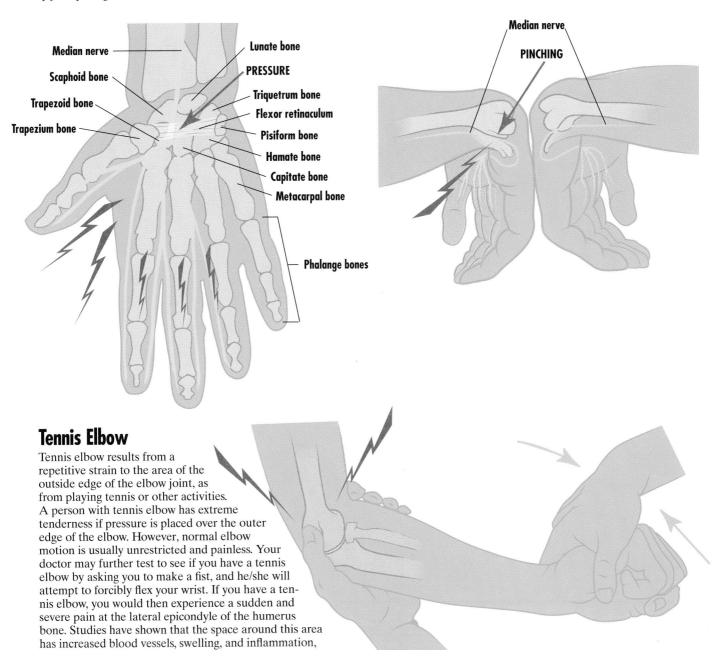

Tennis Elbow

Tennis elbow results from a repetitive strain to the area of the outside edge of the elbow joint, as from playing tennis or other activities. A person with tennis elbow has extreme tenderness if pressure is placed over the outer edge of the elbow. However, normal elbow motion is usually unrestricted and painless. Your doctor may further test to see if you have a tennis elbow by asking you to make a fist, and he/she will attempt to forcibly flex your wrist. If you have a tennis elbow, you would then experience a sudden and severe pain at the lateral epicondyle of the humerus bone. Studies have shown that the space around this area has increased blood vessels, swelling, and inflammation, but no actual tears of the ligaments or muscles.

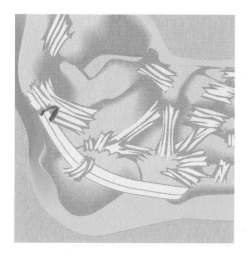

The Legs and Feet

YOUR LOWER EXTREMITY is composed of the thigh, the leg, and the foot. The main joints include the hips, knees, and ankles. Your lower extremity's main functions are to allow weight bearing and walking. This is accomplished by the ankle and the foot, which are the focal points by which the total weight of your body is transmitted into ambulation. Additionally, your foot has a thick heel and toe pads that act as shock absorbers for walking and running. As a child, first you learn to crawl and then to walk. Later, your legs and feet are able to transport you from place to place. It is from this time that you can use your legs and feet to do many useful and pleasurable things—kick a football, ski down a mountain, or run to the lottery office to pick up your million-dollar check.

The bones of your lower extremity include the femur, fibula, tibia, patella, and the foot bones. The femur, or thigh bone, is the longest, largest, and heaviest bone of the body. The tibia, in your leg, is the second largest bone in your body. Also in your leg are the fibula bone, the most slender of all your long bones, and the patella, which is a flat, triangular-shaped bone found in the front of the knee and within the tendon of the quadriceps femoris muscle. The bones of the ankle and foot contain seven tarsal bones (some of which are connected to the tibia and fibula bones), five metatarsal bones, and fourteen phalange bones (two for each big toe and three for each of the other toes).

The hip joint is a ball and socket joint that allows you to flex or extend your thigh, moves your thigh toward or away from your body, and rotates your thigh in or out. The joint is formed by the head of the femur bone fitting into a cup-shaped cavity of the pelvis called the acetabulum. The hip bones consist of the ilium, ischium, and pubis bones. They form a bony ring that supports the vertebrae and rests upon the femur bone of the thigh.

The knee joint is composed of two condylar (knucklelike) joints that occur between the femur and tibia bones and a saddle type joint between the femur and the patella bones. The basic movements include flexion, extension, and slight inward and outward rotation. The knee joint has a number of ligaments that hold it together. Some of the more important ones include the anterior and posterior cruciate ligaments and the fibular collateral and tibial collateral ligaments. Two cartilages, the lateral and medial menisci (one on the inside and one on the outside of the knee), help to stabilize the knee, allow the joint to slide, roll, and spin, and assist in lubrication and cushioning of the joint. Additionally, the knee has four bursae, saclike cavities filled with fluid that help prevent friction. One is located in the front, one on the inner surface, one on the outer surface, and one in the back. Since the knee (along with the ankle and foot) supports the entire weight of the body, it is particularly vulnerable to injury involving the bursae, ligaments, tendons, and muscles, and other disorders.

The ankle joint is primarily a hinge joint, which permits movements of extension and flexion. However, the ankle joint does not have pure hinge joint qualities, as it also allows slight amounts of gliding, rotation, and movement of the foot toward or away from the body. The hand and foot joints are constructed somewhat similarly. The bones of the ankle or wrist joint consist of cube-shaped bones (carpal or wrist bones and tarsal or ankle bones), which allow a gliding type of movement. The middle group of bones (metacarpal or hand bones, and metatarsal or foot bones) assist the wrist and ankle bones in absorbing the great amount of force put upon them by the body. The toe or finger portions (phalanges) are the most moveable and allow flexion and extension of those parts. However, the hands and feet do differ in that the fingers are much more moveable than the toes, and the bones of the feet are much more solid and less moveable than the bones of the hand.

The great saphenous vein, the largest vein in the body, begins on the inner arch of the foot and goes through the thigh. Within this system of veins and other veins are valves that sometimes become overstretched from increased pressure, as could occur in pregnancy or prolonged standing, which may then result in varicose veins. Varicose veins typically occur in the leg and result from a loss of normal valve activity and abnormal dilation of a vein or veins, especially in the great saphenous vein system (also see Chapter 10). Varicose veins occur more frequently in the legs than in, say, the arms because of the large amount of pressure and weight exerted on the leg valves and veins from the effects of gravity and the distance that the blood must flow from the heart to the leg. The leg veins of a person with varicose veins often appear dark blue and snakelike, and they frequently produce an aching in the leg and swelling of the ankle at the end of the day.

Throughout life there are many conditions that can change the structure and function of the hip and thereby speed up the normal wear and tear on the hip joint. These conditions include osteoporosis, arthritis, and injuries like hip fractures. A typical person who gets a hip fracture is a woman over the age of 65. The reason may be that women are much more likely to develop osteoporosis, which weakens the bones and increases the chance that a bone will fracture if there is a fall. In some instances, the brittle bone itself can collapse, causing the fall rather than the fall causing the fracture. It is estimated that the hip fracture rate doubles every five years after age 50, and that women have two to four times as many hip fractures as men.

Sprains of the knee ligaments are very common in athletes. The usual cause of a mild to moderate knee sprain is an impact to the back outside portion of the knee while the foot is solidly on the ground. These injuries can result in tenderness, swelling of the injured tissue, a small amount of bleeding in the joint, and a joint that may be more moveable than it should be. Severe sprains, with complete rupture of a ligament, result in the above damage plus an unstable joint. A more severe injury, such as a rupture of an anterior cruciate ligament of the knee, is another that frequently occurs during rough athletic games—when a basketball player twists his extended knee during landing after a jump shot, for example. Often the player hears a pop and feels a strong pain in the knee. Rupture of this anterior cruciate ligament produces a knee that can be very unstable when the person tries to stand on it. The knee "gives way" and may slip or slide, especially when the person tries to turn to the right or left when the foot is still on the ground. This instability is the result of the tibia bone frontally moving out of its normal position on the femur. Another knee ligament, the posterior cruciate ligament, is frequently ruptured during severe extension of the knee or by a blow hitting the front portion of a flexed knee. When the posterior cruciate ligament ruptures, the tibia moves backward out of its normal position, and the front surface of the leg appears to sag. The knee is excessively extended and the joint has more range of movement than it should, causing an unstable knee.

Other knee problems include bursitis, tendinitis, stress fractures, and injuries caused by overuse of the joint. Bursitis of the knee produces an inflammation of a bursa between the patella and the skin. Usually the cause is not known, but it can result from repeated friction to the knee, as can occur in maids or nuns who do repeated kneeling. It can also result from an infection, injury, or diseases such as arthritis or gout.

Tendinitis of the knee causes an inflammation of any tendon in the area. It can result from trauma or overuse of that particular muscle, and may, in some cases, cause calcification and hardening at the knee.

Stress fractures can occur at any age, and result in a partial break in a cortical area of the bone. They are usually related to a repetitive and prolonged action or stress to that bone. If the activity is not discontinued, a full fracture can occur. Running is the most common cause, and in the knee area it usually affects the tibia or fibula bones. Stress fractures in either of these two bones frequently produce pain, swelling, and warmth over the stress fracture site.

Knees, as well as other joints, may develop a disorder called overuse synovitis. Synovitis is an inflammation of the synovial membranes (which line the joint capsule and secrete a lubricating fluid), and overuse synovitis occurs from an increased and perceived overuse of a body part that has these synovial membranes. The synovial membrane of the knee joint is the largest and most extensive of the body. Overuse synovitis frequently occurs in runners, particularly those who rapidly increase their mileage.

The most frequent injury to the ankle is a sprain. A sprain is a stretching or tearing of a ligament by a motion beyond its normal range. If the sprain is mild, then no torn ligaments are found. However, if it is severe, there may be tearing of ligaments, such as between the fibula and the talus (a foot bone).

The following illustrations will give you a better understanding into the legs and feet and some of their common disorders.

The Lower Extremity, Hip, and Knee

The Lower Extremity

These figures show some of the important muscles, arteries, bones, and nerves of the front and back of the lower extremity.

Sartorius muscle—The longest muscle in the body, and is narrow and ribbon-like. It flexes the thigh, and rotates it inward or outward.

Tensor fascia latae—Flexes the thigh, and rotates the thigh inward.

Rectus femoris muscle—Part of a muscle group that extends the thigh and leg.

Vastus lateralis muscle—Extends the thigh and the leg.

Quadriceps femoris muscle tendon—The tendon of the quadriceps femoris muscle, which extends the thigh and the leg.

Biceps femoris muscle tendon—Flexes the leg, and after it is flexed, it rotates it outward.

Peroneus longus muscle—Flexes the foot downward and lowers the inner part of the foot toward the ground.

Tibialis anterior muscle—Flexes the foot up and inverts it.

Extensor retinaculum

Peroneus tertius muscle tendon—Flexes the foot up, and lowers and twists the inner part of the foot so that it approaches the ground.

Iliopsoas muscle—Flexes the thigh and flexes the lumbar vertebral column to either side.

Adductor longus muscle—Flexes and rotates the thigh, and moves the thigh inward.

Adductor magnus muscle—The upper part rotates the thigh inward, and flexes it. The lower part extends and rotates the thigh outward.

Vastus medialis muscle—Extends the thigh and the leg.

Knee joint capsule

Soleus muscle—Flexes the foot down.

Extensor digitorum longus muscle tendons—Extends the phalanges of the four toes (other than the big toe), flexes the foot up and lowers and twists the inner portion of the foot so that it approaches the ground.

Extensor hallucis longus muscle tendon—Extends the big toe, and flexes the foot up.

Gluteus medius muscle—Draws the thigh out and rotates it inwards.

Gluteus maximus muscle—Extends and rotates the thigh outwards.

Semitendinosus muscle—Flexes the leg and rotates it inward, and extends the thigh.

Semimembranous muscle—Flexes the leg and rotates it inward, and extends the thigh.

Plantaris muscle—Flexes the foot down, and flexes the leg.

Gastrocnemius muscle—Flexes the foot down, and flexes the leg.

Achilles tendon

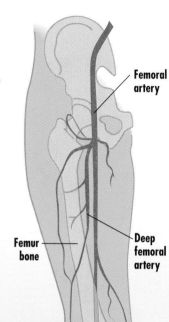

Femoral artery

Deep femoral artery

Femur bone

Patella bone

Posterior tibial artery

Fibula bone

Tibia bone

Peroneal artery

Tarsal bones

Metatarsal bones

Phalange bones

The Hip Joint

This figure depicts the hip joint. Your hip joint allows you the capability to flex or extend your thigh, move your thigh toward or away from your body, and rotate your thigh in or out.

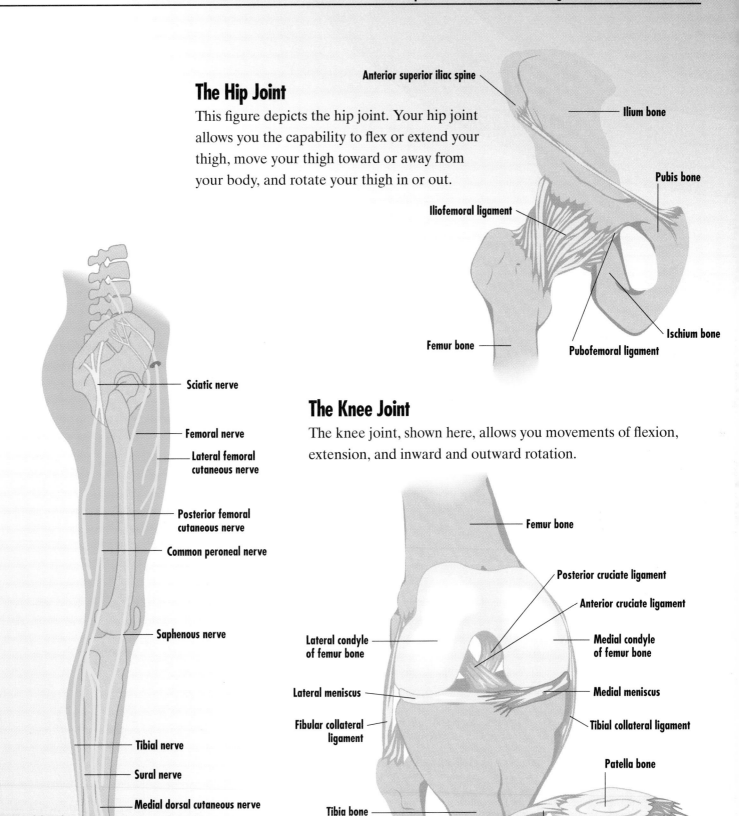

Anterior superior iliac spine

Ilium bone

Pubis bone

Iliofemoral ligament

Femur bone

Pubofemoral ligament

Ischium bone

Sciatic nerve

Femoral nerve

Lateral femoral cutaneous nerve

Posterior femoral cutaneous nerve

Common peroneal nerve

Saphenous nerve

Tibial nerve

Sural nerve

Medial dorsal cutaneous nerve

Lateral dorsal cutaneous nerve

Lateral plantar nerve

Medial plantar nerve

The Knee Joint

The knee joint, shown here, allows you movements of flexion, extension, and inward and outward rotation.

Femur bone

Posterior cruciate ligament

Anterior cruciate ligament

Lateral condyle of femur bone

Medial condyle of femur bone

Lateral meniscus

Medial meniscus

Fibular collateral ligament

Tibial collateral ligament

Patella bone

Tibia bone

Fibula bone

Patellar ligament

Hip Fracture, Knee Ligament Injuries, and Ankle Sprain

Hip Fracture

Hip fractures can occur in anyone as a result of an injury, but are more common in older women. There are about 250,000 hip fractures per year, and over the past ten years the number of hip fractures has increased. The causes include trauma, injuries, osteoporosis, and old age with a loss of strength in the trabecular bone. However, fractures may also be due to falls resulting from dizziness, vertigo, or stability problems, and from changes in the vestibular system of the ears, heart problems, adverse side effects of medicines, alcoholism, or other health concerns.

This figure shows a fracture of the neck of the femur bone. These types of fractures are usually associated with falls, and are common in the elderly.

— **Fracture of the neck of the femur bone**

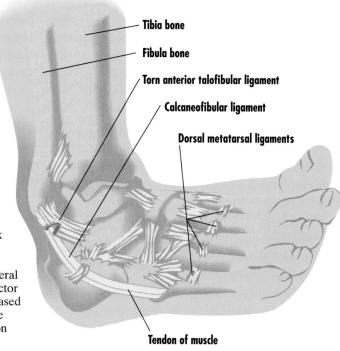

— **Tibia bone**

— **Fibula bone**

— **Torn anterior talofibular ligament**

— **Calcaneofibular ligament**

— **Dorsal metatarsal ligaments**

Tendon of muscle

Ankle Sprain

An ankle sprain can often occur from an injury that involves an inversion of the ankle, with stretching or tearing of a ligament. The anterior talofibular ligament and calcaneofibular ligaments are weak and easily torn, and are usually involved in ankle sprains. A person with a moderate to severe ankle sprain usually has severe pain, swelling from bleeding, and black and blue marks that develop several hours later. To determine if the person has a sprained ankle, the doctor may flex the foot downward toward the ground, and invert it. Increased pain indicates that the ligament may be sprained or torn. The figure here shows some of the ligaments of the ankle and the most common site of ankle sprains, the anterior talofibular ligament.

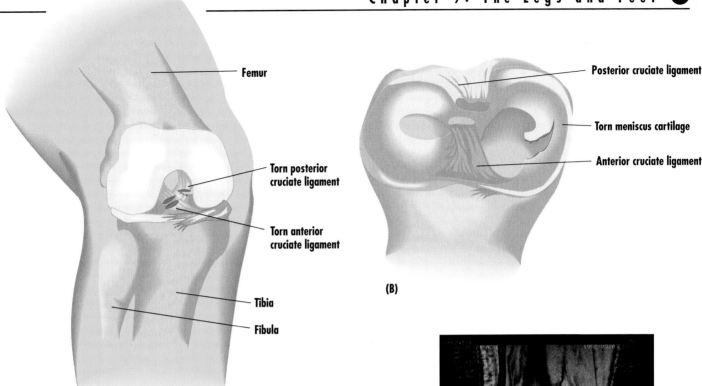

Femur

Torn posterior cruciate ligament

Torn anterior cruciate ligament

Tibia

Fibula

(A)

Posterior cruciate ligament

Torn meniscus cartilage

Anterior cruciate ligament

(B)

Knee Ligament Injuries

Knee ligament injuries are very common in people who participate in contact sports like football or other sports or activities that place stress on the joints, such as basketball. The injuries can range from minimal stretching to complete tearing of the ligaments. In severe cases, people can have extensive pain, swelling, looseness of the joint, and discoloration of the skin of the knee from internal bleeding in that area. The anterior cruciate ligament (Figure A) is one of the knee ligaments that can be frequently torn during strenuous exercise. If this occurs, the injured person complains that his or her knee gives way and slips or slides with a turn to the right or the left. The posterior cruciate ligament (Figure A) is the main stabilizer of the knee for extension. The most frequent causes of posterior cruciate ligament tears are too much extension of the knee or a blow to the front of a flexed knee. When there is a rupture of this ligament, the tibia bone moves backward, the front surface of the leg seems to sag, and the knee joint is unstable.

Tearing of the meniscal cartilage of the knee is another common injury. The menisci are two c-shaped cartilages, one on the inner part of the joint and one on the outer part of the joint (Figure B). They cushion the knee and are frequently injured as a result of a twisting injury to one or both menisci. People who suffer a meniscal tear develop pain at the knee joint and have an inability to flex or extend the knee properly.

Figure C shows an MRI of a normal knee. You can see the bow-tie shaped appearance of the menisci, which are normal. However, Figure D shows a torn meniscus of the right knee, and here, you can see that the bow-tie shaped appearance is now deranged.

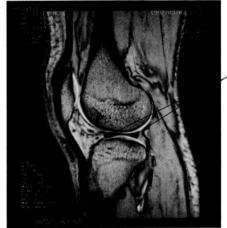

Normal menisci

(C) Normal menisci of the knee
Image courtesy of GE Medical Systems

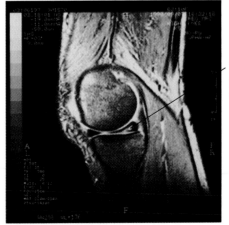

Torn meniscal cartilage

(D) Torn meniscal cartilage of the right knee
Image courtesy of GE Medical Systems

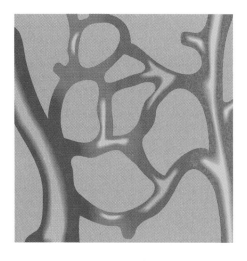

CHAPTER
10

The Heart and Circulation

YOUR CARDIOVASCULAR SYSTEM, or circulatory system, consists of the heart, the blood, and a series of blood vessels and lymphatic vessels. The purposes of the circulatory system are to deliver oxygen and nutrients to all parts of your body and also to pick up waste materials and poisons that the body will subsequently eliminate. There are various types of blood vessels, including the arteries, arterioles (smaller arteries), veins, venules (smaller veins), and capillaries. These vessels participate in what are generally considered to be two systems within the circulatory system. The first one, called the systemic circulatory system, carries oxygen-rich blood from the heart to the organs and tissues. The second one, called the pulmonary circulatory system, transports oxygen-poor blood and other materials from the organs and tissues to the lungs, where carbon dioxide is released and oxygen is picked up, and then carries the oxygen-rich blood back to the heart.

Blood is an important component of your circulatory system. It is essential not only in transporting oxygen, nutrients, and wastes but also in keeping your temperature stable, helping your body fight against infections and allergies, and helping your skin heal by forming a blood clot after you have cut yourself. Blood makes up about 7% of your body weight. The actual amount of blood varies from person to person, but an average-sized man has about 5 liters of blood in his body. The blood is channeled in the body through two primary types of vessels, the arteries and the veins.

The arteries are thick-walled vessels that usually carry blood with a high concentration of oxygen and a low concentration of carbon dioxide. Under high pressure, the arteries move this blood from the heart to the organs and tissues. The aorta, which arises directly from the heart, is the largest artery in the body and carries blood away from the heart. The coronary arteries, which arise directly from the beginning of the aorta, are important because they are the vessels that deliver the blood supply to the heart and therefore directly nourish the heart. Diseases that affect the coronary arteries can lead to chest pain (angina) and heart attack (myocardial infarction).

The veins are thin-walled vessels. While almost all of the veins carry blood with a lower concentration of oxygen and a high concentration of carbon dioxide, there is one exception: the pulmonary veins, which carry blood with a high concentration of oxygen. Two of the largest veins in your body are the superior vena cava and the inferior vena cava, which enter directly into the heart and carry blood that has come from the tissues and organs. Many veins have valves that prevent a backflow of blood. If the veins are placed under excessive or prolonged pressure, they can become overstretched and the valves can become destroyed or incompetent. The condition that results, varicose veins, is usually accompanied by swelling of the legs.

Now let's explore the route of circulation. After oxygen-rich blood leaves your heart via the aorta, it goes to other arteries, then to the arterioles, and on to the capillaries, which bring nourishment directly to the tissues. Capillaries are the smallest blood vessels, and there are approximately 10 billion of them in your body. They have extremely thin, permeable walls that allow for the transfer of substances to and from the tissues. While one group of capillaries delivers nutritious substances to the tissues, another group takes up the waste materials and ships them out of the tissues. Some of these respiratory wastes will go to the liver and kidneys and be excreted through the urine and the stool. Other respiratory wastes will go back through the circulation and will be excreted through the lungs. (Chapters 12 and 13 discuss how the body's other waste materials are disposed of.) The blood that contains the waste materials is then channeled through the venules, to the veins, to the superior and inferior vena cava, and on to the heart, where it is taken up by the pulmonary arteries. The blood then travels via these pulmonary arteries to the lungs, where it releases carbon dioxide and other respiratory wastes through expiration and also receives new oxygen. This oxygen-rich blood then travels via the pulmonary veins to the heart, and then to the aorta, at which point the whole process is repeated.

Your heart is composed primarily of muscles that together act as a pump. The heart has four compartments: two atria, which are the upper chambers, and two ventricles, which are the lower chambers. A piece of tissue, called the septum, divides the left atrium and left ventricle from the right atrium and right ventricle. The blood that enters the atrium and ventricle on the right side of the heart is oxygen-poor blood. This blood moves from the right atrium to the right ventricle and then goes via the pulmonary arteries to the lungs, where it picks up oxygen. Now the blood, which is oxygen-rich, moves via the pulmonary veins to the left atrium and then to the left ventricle. Each chamber has a valve that allows passage of the blood into the next chamber. There are two different types of valves: atrioventricular (AV) and semilunar. The tricuspid valve (the one between the right atrium and right ventricle) and the mitral valve (the one between the left atrium and left ventricle) are AV valves. The pulmonary valve (the one leaving the right ventricle) and the aortic valve (the one leaving the left ventricle) are semilunar valves.

Your heartbeat is sustained by a part of the heart that functions as a pacemaker and is called the sinoatrial (SA) node. The healthy heart has a regular beat. While a rate between 60 and 100 beats per minute is considered normal, people's heart rates vary depending on factors such as age, sex, physical activity, and emotions. The pulses of women and children are usually slightly higher than those of men, with children's pulses ranging from 90 to 120. Well-trained athletes often have pulses in the 50s or lower. Your pulse can increase with fevers, bleeding, anemia, heart disease, or thyroid disease. It can decrease with fainting or with certain heart, liver, or brain problems.

When your doctor listens to your heart with a stethoscope, he or she is hearing vibrations from the heart that are brief and of very low frequencies. The first heart sound, or "lubb," is associated with the contraction of the heart muscles, the closing of the AV valves, the subsequent development of pressure, and the opening of the semilunar valves. The loudness depends on the rigor of the ventricles contracting and the stiffness of the position of the mitral valve. The second heart sound, or "dubb," is caused by the tensing or closing of the semilunar valves and by vibration of the valves, heart, and large arteries. The second heart sound is normally of a higher frequency and shorter duration than the first heart sound. The third and fourth heart sounds are usually not heard, and a heart problem may exist if they can be heard with a stethoscope.

In addition to checking your blood pressure, measuring your pulse, and listening to your heart with a stethoscope, your doctor may wish to pursue other tests that help in the evaluation of the heart and circulatory system. These include electrocardiograms (EKGs), echocardiograms, stress tests, and thallium stress tests to help assess your risk of a heart attack. As with most tests, none of the ones listed are 100% accurate in determining heart disease, but each can provide valuable information.

Heart attacks, or myocardial infarctions, are one of the most common and serious diseases. Each year, about 1.5 million people in the United States suffer from a heart attack, and about 25% of all deaths result from this. The majority of heart attacks are attributable to arteriosclerosis, which is commonly called hardening of the arteries. In the following pages, you will learn more about heart attacks, high blood pressure, and other common heart problems.

The Circulation and the Heart

The Circulatory System

This diagram shows some of the major arteries and veins that leave and enter your heart.

A Magnified View of Blood Vessels

Diagram A shows a vein with a valve. Valves help prevent the backflow of blood in the wrong direction.

Diagram B shows a representation of an artery, which has the thickest walls of all the blood vessels.

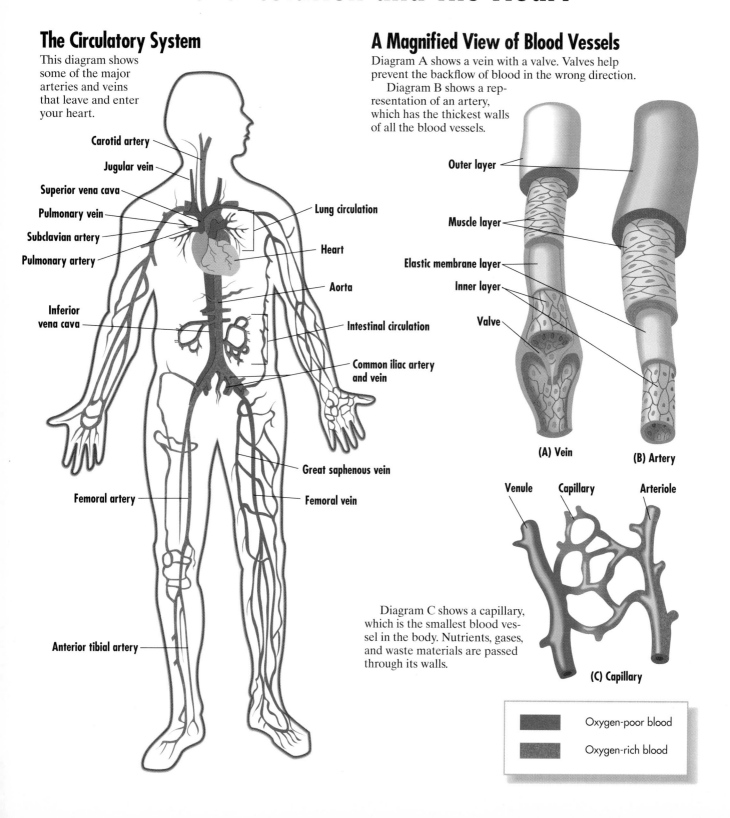

Carotid artery

Jugular vein

Superior vena cava

Pulmonary vein

Subclavian artery

Pulmonary artery

Inferior vena cava

Lung circulation

Heart

Aorta

Intestinal circulation

Common iliac artery and vein

Great saphenous vein

Femoral artery

Femoral vein

Anterior tibial artery

Outer layer

Muscle layer

Elastic membrane layer

Inner layer

Valve

(A) Vein

(B) Artery

Venule

Capillary

Arteriole

Diagram C shows a capillary, which is the smallest blood vessel in the body. Nutrients, gases, and waste materials are passed through its walls.

(C) Capillary

Oxygen-poor blood

Oxygen-rich blood

The Heart and the Coronary Arteries

The coronary arteries are extremely important, since they supply nourishment to the heart. The main coronary arteries lie on the surface of the heart, and smaller arteries traverse directly into the heart muscles. The left coronary artery predominantly nourishes the left ventricle, while the right coronary artery mainly supplies blood to the right ventricle, with a small portion going to the left ventricle. If arteriosclerosis (hardening and narrowing of arteries) develops, it frequently affects the coronary arteries.

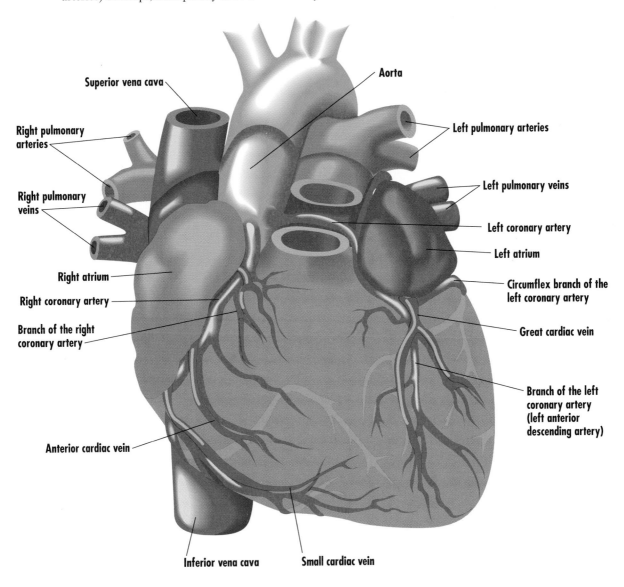

Superior vena cava

Aorta

Right pulmonary arteries

Left pulmonary arteries

Right pulmonary veins

Left pulmonary veins

Left coronary artery

Left atrium

Right atrium

Right coronary artery

Circumflex branch of the left coronary artery

Branch of the right coronary artery

Great cardiac vein

Branch of the left coronary artery (left anterior descending artery)

Anterior cardiac vein

Inferior vena cava

Small cardiac vein

Blood Cells

This illustration shows examples of all the types of cells that we have in our blood. The neutrophils, bands, eosinophils, monocytes, lymphocytes, and basophils are all types of white blood cells (WBCs). The arrangement of the cells in this figure is random, and the number of white blood cells in relation to the number of red blood cells (RBCs) and platelets is greater than what would actually occur. Plasma, the fluid portion of the blood, carries all of the above types of blood cells, as well as nutrients and additional materials, from one part of the body to another.

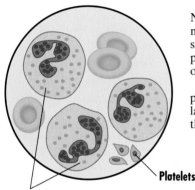

Neutrophils, the most common WBC, are twice the size of RBCs, and are important in ridding the body of infection.

The main function of platelets is to help coagulate (clot) the blood when this is necessary.

Platelets

Neutrophils

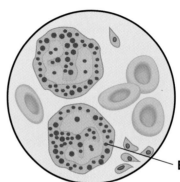

Basophils help maintain the fluidity of the blood and are important in allergic reactions. While some basophils are in the blood, others are in tissues. They can be increased with some cancers, infections, allergies, or radiation exposure.

Basophil

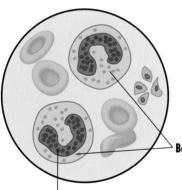

Bands are immature WBCs. They have a horseshoe appearance to their nucleus, and the number of bands is increased in the presence of infection, since bands also help fight infection.

Band cells

Nucleus of band cell

Lymphocyte

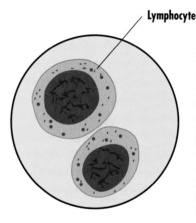

Lymphocytes are predominantly found in the lymph, which is a slightly yellow liquid derived from the tissue fluids. Lymphocytes are the second most common WBC of the blood and are important for the proper functioning of the immune system. Their numbers can increase with infections, some glandular problems, leukemias, and certain intestinal problems, such as Crohn's disease.

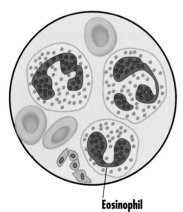

Eosinophils are a type of WBC that have large spherical granules that stain red. The number of eosinophils is increased in the presence of parasites, skin diseases, asthma, hay fever, other allergies, or some mild infections, whereas the number is decreased in the presence of stress, severe infections, burns, shock, or steroid use.

Eosinophil

Monocyte

Cytoplasm of monocyte

Nucleus of monocyte

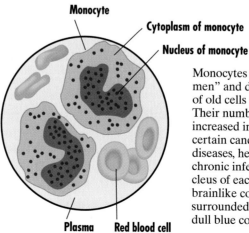

Monocytes act like "PAC-men" and devour fragments of old cells and cancer cells. Their numbers and size can be increased in the presence of certain cancers, autoimmune diseases, heart infections, or chronic infections. The nucleus of each monocyte has brainlike convolutions and is surrounded by cytoplasm of a dull blue color.

Plasma   **Red blood cell**

Anemia

Iron deficiency anemia is a common blood disorder and occurs most frequently during two periods of life: infancy and pregnancy. Although iron deficiency anemia is often not serious, it may be present as a warning sign of cancer. Pernicious anemia is another blood disorder. It is caused by the inability of the body to properly absorb vitamin B_{12} because of the lack of a substance called intrinsic factor.

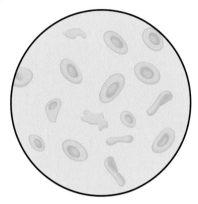

Normal Red Blood Cells

The RBCs have the main functions of transporting oxygen to the tissues and of returning carbon dioxide to the lungs from the tissues. They are helped by an oxygen-carrying protein, hemoglobin. The RBCs also contain enzymes for energy and help maintain the acidity and alkalinity of the body.

Red Blood Cells in Iron Deficiency Anemia

Red blood cells in iron deficiency anemia are smaller and paler than normal cells, with the paleness due to a decrease in hemoglobin. People with this type of anemia may have a red tongue, spoon-shaped nails, and an enlarged spleen.

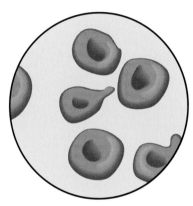

Red Blood Cells in Pernicious Anemia

Red blood cells in pernicious anemia are large and usually round. The four most common symptoms of this type of anemia are fatigue, sore tongue, short-ness of breath, and abnormal feelings in the hands and feet.

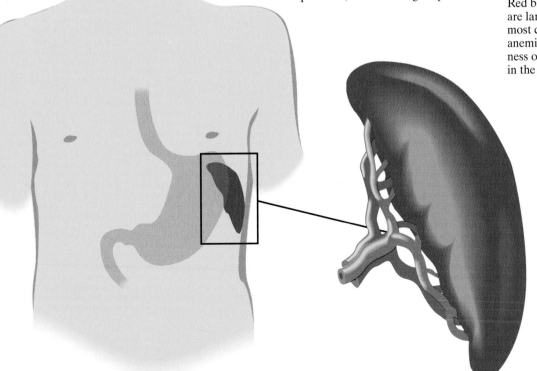

Spleen

The spleen is the largest organ composed of lymph tissue. It is approximately the size of a fist and is located in the upper-left part of the abdomen, behind the stomach. It is dark red and spongy. It produces and stores blood cells and filters debris. During times of stress or exercise, the spleen can contract and send more blood into the rest of the body. It can become enlarged in illnesses such as alcoholism, mononucleosis, tuberculosis, malaria, and some cancers.

How the Heart Works

These illustrations show two major, and very connected, activities of the heart: how blood flows through the heart and how the heart beats. The heart is predominantly composed of muscles that act as a pump for circulating blood. Your heart also contains nerves that affect the pumping of the heart by changing the rate of the heartbeat and the strength of the heart's contraction. The heartbeat is a function of nerves and electrical impulses that work together to contract heart muscles and to open and close valves.

The Flow of Blood through the Heart

2 The blood then passes through the right atrium, right ventricle, and pulmonary arteries on its way to the lungs.

1 Blood enters the right side of the heart from the superior and inferior vena cava.

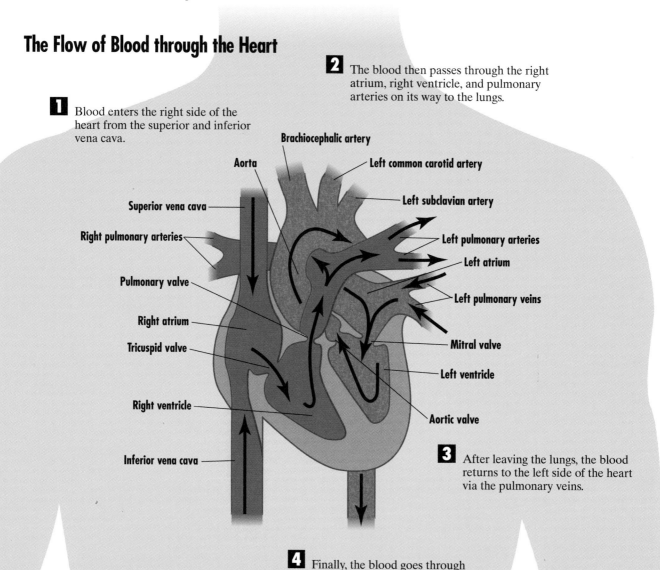

Brachiocephalic artery

Aorta

Left common carotid artery

Superior vena cava

Left subclavian artery

Right pulmonary arteries

Left pulmonary arteries

Left atrium

Pulmonary valve

Left pulmonary veins

Right atrium

Tricuspid valve

Mitral valve

Left ventricle

Right ventricle

Aortic valve

Inferior vena cava

3 After leaving the lungs, the blood returns to the left side of the heart via the pulmonary veins.

4 Finally, the blood goes through the left atrium and left ventricle, and passes through the aorta on its way to the rest of the body.

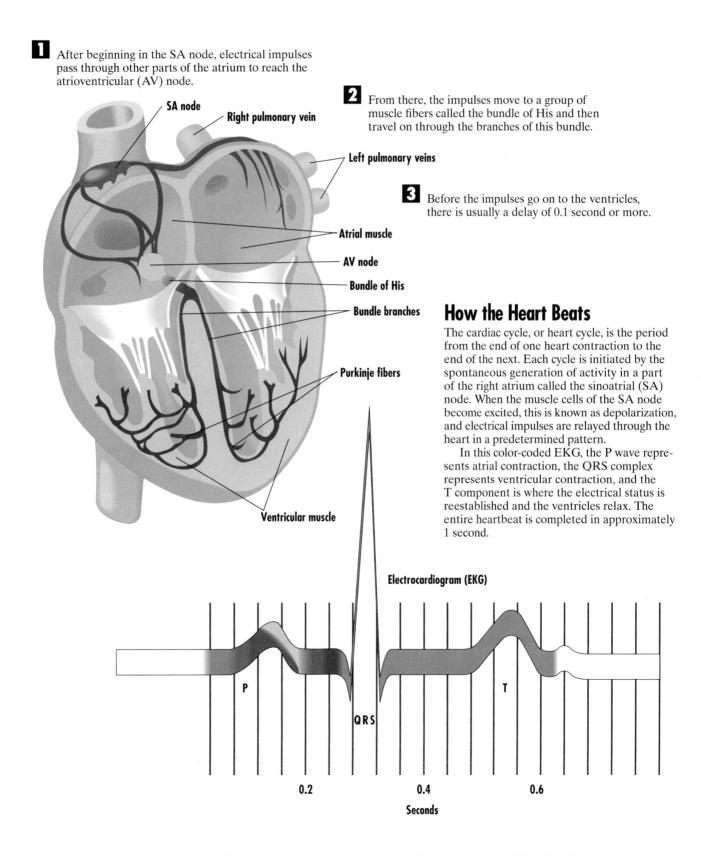

1 After beginning in the SA node, electrical impulses pass through other parts of the atrium to reach the atrioventricular (AV) node.

2 From there, the impulses move to a group of muscle fibers called the bundle of His and then travel on through the branches of this bundle.

3 Before the impulses go on to the ventricles, there is usually a delay of 0.1 second or more.

SA node

Right pulmonary vein

Left pulmonary veins

Atrial muscle

AV node

Bundle of His

Bundle branches

Purkinje fibers

Ventricular muscle

How the Heart Beats

The cardiac cycle, or heart cycle, is the period from the end of one heart contraction to the end of the next. Each cycle is initiated by the spontaneous generation of activity in a part of the right atrium called the sinoatrial (SA) node. When the muscle cells of the SA node become excited, this is known as depolarization, and electrical impulses are relayed through the heart in a predetermined pattern.

In this color-coded EKG, the P wave represents atrial contraction, the QRS complex represents ventricular contraction, and the T component is where the electrical status is reestablished and the ventricles relax. The entire heartbeat is completed in approximately 1 second.

Electrocardiogram (EKG)

P

QRS

T

0.2

0.4

0.6

Seconds

Hardening of the Arteries, High Blood Pressure, and Heart Attacks

Hardening of the arteries (arteriosclerosis), high blood pressure, and heart attacks (myocardial infarctions) are the most common disorders of the heart and circulatory system. Many doctors believe that coronary artery disease can occur as a result of high blood pressure, high cholesterol levels, or a combination of the two. Cholesterol is a fatlike substance that is essential for life, since it is an important component of hormones and helps form a coating for nerves and cells. However, most experts believe that the risk for a heart attack is increased if the level of cholesterol in the blood exceeds 200 milligrams per deciliter.

When blood pressure is measured, the results are usually expressed in two numbers (for example, 120 over 80). The first number, which is the higher of the two, is the systolic pressure, while the second number is the diastolic pressure. Systole is the period of contraction of the ventricles. It is followed by diastole, which is the period of relaxation, or the period when the heart refills. A normal systolic blood pressure is less than 140, while a normal diastolic blood pressure is less than 90.

Almost all heart attacks are due to narrowing or hardening of the coronary arteries (arteriosclerosis) and to the presence of one or more blood clots in the heart (coronary thrombosis). Recent studies have confirmed that emotional stress and the time and severity of physical activity can be potential triggers for heart attacks, although routine exercise, in the long run, generally improves people's health.

Hardening of the Arteries

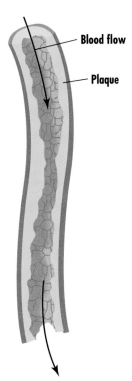

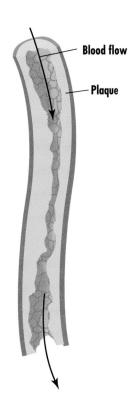

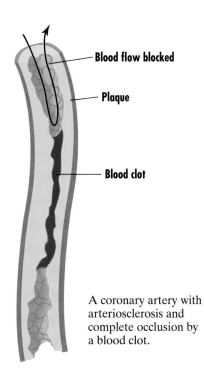

A coronary artery with arteriosclerosis and complete occlusion by a blood clot.

A moderate amount of arteriosclerosis with narrowing of the coronary artery vessel.

Arteriosclerosis of a coronary artery with deposits (plaques) of calcium or other material. The vessel is greatly occluded. One possible explanation for how this develops is that the vessel is first irritated or injured by exposure to a factor, such as smoking. Then platelets adhere to the injured area and secrete hormones that cause fibrosis, which is a thickening of the tissue. Subsequent accumulation of fat in the injured area may lead to the formation of pearly patches called plaques and may result in hardening, or calcification, of the blood vessel.

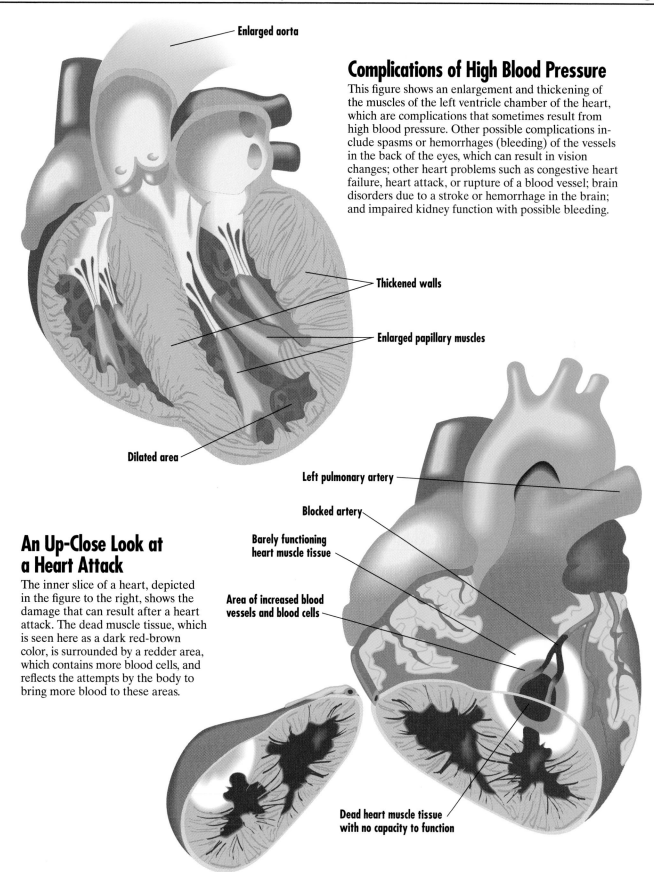

Enlarged aorta

Complications of High Blood Pressure

This figure shows an enlargement and thickening of the muscles of the left ventricle chamber of the heart, which are complications that sometimes result from high blood pressure. Other possible complications include spasms or hemorrhages (bleeding) of the vessels in the back of the eyes, which can result in vision changes; other heart problems such as congestive heart failure, heart attack, or rupture of a blood vessel; brain disorders due to a stroke or hemorrhage in the brain; and impaired kidney function with possible bleeding.

Thickened walls

Enlarged papillary muscles

Dilated area

An Up-Close Look at a Heart Attack

The inner slice of a heart, depicted in the figure to the right, shows the damage that can result after a heart attack. The dead muscle tissue, which is seen here as a dark red-brown color, is surrounded by a redder area, which contains more blood cells, and reflects the attempts by the body to bring more blood to these areas.

Left pulmonary artery

Blocked artery

Barely functioning heart muscle tissue

Area of increased blood vessels and blood cells

Dead heart muscle tissue with no capacity to function

The Lungs

YOUR BREASTBONE, RIBS, and spine provide good protection for your two lungs, which function primarily in breathing and filtering air. The atmospheric air that you breathe consists of 78–79% nitrogen, 20–21% oxygen, 0.5% water, 0.05% carbon dioxide, and other particles. This air enters via the nose and mouth, travels through the larynx (voice box), and proceeds down the trachea (windpipe), which branches into the two main bronchi. From here, the air is channeled into bronchioles (smaller bronchi) and ultimately into the millions of alveoli (air sacs) that are in the lungs. Each air sac has extremely thin walls that contain a meshwork of capillaries, which are involved in the exchange of oxygen and carbon dioxide. This exchange of gases in the blood is referred to as ventilation.

Respiration, or breathing, takes place from 10 to 15 times per minute. Although you can control it consciously when you wish, your brain usually controls it unconsciously, with the assistance of the respiratory center of the brain (see Chapter 1). Messages are sent from this center to your diaphragm and to certain rib muscles, which then contract. This contraction pulls the lower surfaces of the lungs downward, and you inhale the air. Stretch receptors in the lung then send signals back to the brain. These signals make the diaphragm and certain rib muscles relax, which causes an upward movement of the diaphragm, and you exhale the air.

The respiratory muscles in your chest assist in breathing and ventilation. They do this by compressing and distending the lungs and causing an expansion and contraction of the alveoli. The lungs are elastic in nature, and this facilitates ventilation, as does the presence of surfactant, a material found in the lungs of healthy individuals. The surfactant acts similarly to a detergent, decreasing the tension of fluids that lie on the alveoli. If you lacked surfactant, like some people with congenital diseases do, you would have difficulty expanding your lungs.

Each lung is covered by a pleura, which protects the lung and also helps it expand and contract easily inside the chest. The lungs are spongy in consistency, but are also light in weight and can float in water.

Normally, your lungs are able to avoid infections because the nose and respiratory system do an effective job of filtering the air, and because the trachea and bronchi produce mucus, which assists in trapping and carrying away contaminants. Cilia, which are small hairs that cover the entire respiratory tract, move the mucus and contaminants along by rapidly beating back and forth, in some areas at a rate of up to 1000 times per minute.

The respiratory cycle consists of inspiration (breathing in) and expiration (breathing out). The quantity of air you breathe in or out with each normal breath is called your tidal volume. In adults, the tidal volume is usually about 500 milliliters (ml), which is half a liter. The quantity of air you exhale, following your deepest inspiration of air, is called your vital capacity. The vital capacity in adults is about 4600 ml in men and about 3500 ml in women, although higher amounts can be seen in tall and thin people and in well-conditioned athletes. The total volume of both lungs ranges from 5800 to 6500 ml in adults.

The respiratory rate, or breathing rate, can increase with fever, an increase in blood pressure, exercise, or anything that produces more carbon dioxide in the body. The respiratory rate can decrease during sleep or periods of rest, or with a decrease in body temperature, blood pressure, or oxygen in the body.

There are many infections and diseases of the lungs, but we will only touch on a few of them. Asthma is usually a chronic, inflammatory condition that can cause patients to have attacks of wheezing, dry cough, or difficulty in breathing. The attacks may be provoked by emotions, irritants, exercise, infections in the respiratory system, fatigue, or allergies to animals, molds, dust, or other substances. When the patient is exposed to the provoking agent, the overly sensitive bronchi or bronchioles undergo spasms or become narrow. The result is an asthma attack. People with asthma usually have little or no trouble inspiring, but they do have trouble breathing out, or expiring; therefore, to help determine the severity of an asthma attack, doctors or even patients can measure their expiratory flow rate.

Bronchitis and pneumonia, two other lung-related illnesses, are frequently caused by infectious organisms (such as bacteria and viruses) or exposure to irritants (such as chemicals). Bronchitis is an inflammation of the bronchi. Pneumonia can affect the bronchi and the alveoli, causing the alveoli to become filled with fluid and blood cells. Many experts recommend that the elderly, people with chronic diseases, and health care professionals talk to their doctor about obtaining an annual flu shot and a pneumonia vaccine, which can usually be given less often. These vaccines may help prevent the number and severity of flu and pneumonia episodes that are contracted.

Cigarette smoking is known to cause damage to the lungs, bronchi, blood vessels, heart, and other organs and tissues. It can aggravate asthma and can produce bronchitis, cancer, or emphysema. Men who smoke have been reported to have a 70% higher death rate from smoking-related diseases than men who do not smoke, with the main causes of death including lung cancer and coronary heart disease. Smoking is associated with increased risks of all of the following conditions and diseases: coughing, breathing problems, respiratory infections, pneumonia, stroke, hardening of the arteries (arteriosclerosis), stomach and intestinal ulcers, and cancer of the mouth, throat, esophagus, kidneys, bladder, and pancreas. Additionally, smoking during pregnancy increases the risk of miscarriage and fetal death.

More than 4,000 substances have been found in cigarette smoke. Two of the most dangerous substances are nicotine and carbon monoxide. Nicotine is believed to be responsible for the addictive properties of cigarettes. It causes the release of epinephrine, which is a hormone secreted by the body, and this in turn produces an increase in blood pressure and heart rate. Carbon monoxide prevents the blood from carrying the full amount of oxygen, and this is ultimately detrimental to the health.

Secondhand smoke has been reported to increase the risk of respiratory and middle ear infections, especially in children. In the United States, secondhand smoke has also been reported to cause about 3,000 deaths due to lung cancer and about 36,000 deaths due to heart disease each year.

The following illustrations will give you additional information on the functions and diseases of the lungs and the respiratory system.

How the Respiratory System Works

Oxygen is critical in providing your body with energy. Although you might be able to survive for weeks without solid food, your body and especially your brain could not last without oxygen for more than a few minutes. The lungs are the organs that facilitate the intake of oxygen and the elimination of carbon dioxide. When oxygen enters the blood in your lungs, it binds with the hemoglobin of the blood and is then transported to your tissues and organs. Carbon dioxide is also brought to the lungs via capillaries in the air sacs, but it is subsequently removed from the body when you exhale.

Respiration involves a cycle of inspiration, in which you breathe air in, and expiration, in which you breathe air out. During inspiration, the rib cage elevates and expands, and the diaphragm contracts. During expiration, the lungs compress, the diaphragm relaxes, and the rib cage lowers and contracts.

The Respiratory System

The diagram below outlines the pathway of a breath of fresh air. The lungs and respiratory system have a number of components and mechanisms for preventing respiratory infections. These include the epiglottis, which is a flap of tissue that covers the larynx and prevents food particles, bacteria, and other infectious organisms from entering the lungs; the production of mucus, which traps the particles and organisms; the white blood cells, which attack organisms that do reach the alveoli; the cilia, which sweep mucus, trapped organisms, and debris toward the pharynx; and the cough reflex, which expels unwanted materials from the bronchioles.

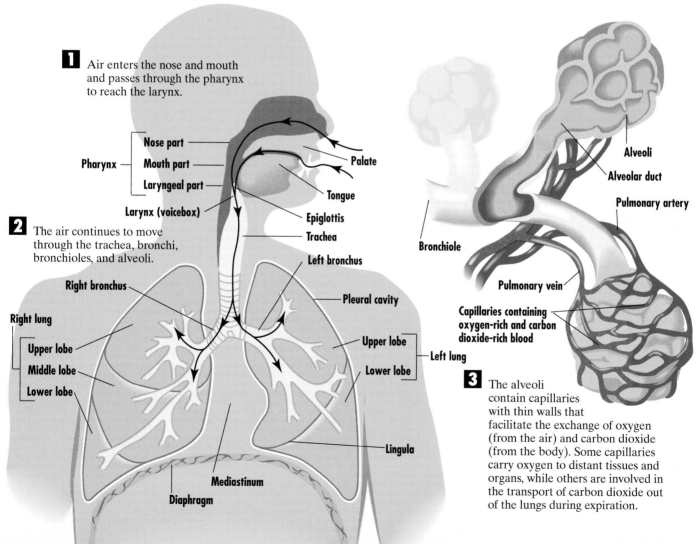

1 Air enters the nose and mouth and passes through the pharynx to reach the larynx.

Nose part
Mouth part
Laryngeal part
Pharynx
Palate
Tongue
Larynx (voicebox)
Epiglottis
Trachea

2 The air continues to move through the trachea, bronchi, bronchioles, and alveoli.

Right bronchus
Left bronchus
Pleural cavity
Right lung
Upper lobe
Middle lobe
Lower lobe
Upper lobe
Left lung
Lower lobe
Lingula
Mediastinum
Diaphragm

Alveoli
Alveolar duct
Pulmonary artery
Bronchiole
Pulmonary vein
Capillaries containing oxygen-rich and carbon dioxide-rich blood

3 The alveoli contain capillaries with thin walls that facilitate the exchange of oxygen (from the air) and carbon dioxide (from the body). Some capillaries carry oxygen to distant tissues and organs, while others are involved in the transport of carbon dioxide out of the lungs during expiration.

A Magnified View of Normal Lung Tissue

This is what a cross-section of healthy lung tissue would look like if magnified. The pleura is the membrane that covers the surface of the lung. A respiratory bronchiole is a terminal branch of a bronchiole. Alveoli arise as outpouchings from the bronchioles; clusters of alveoli are called alveolar sacs.

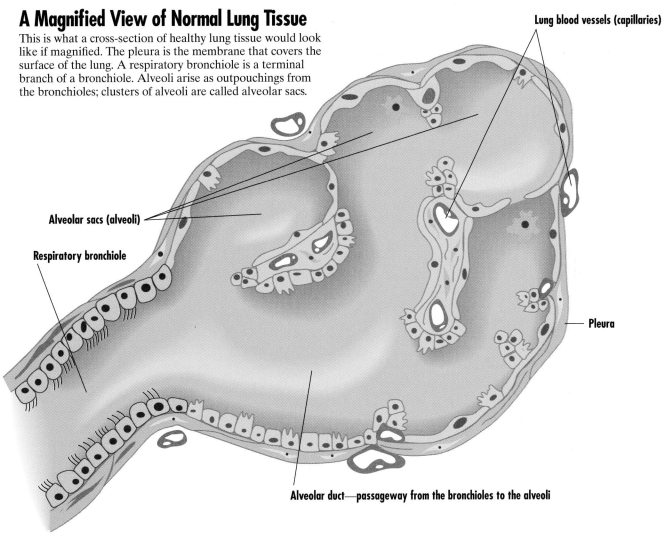

Lung blood vessels (capillaries)

Alveolar sacs (alveoli)

Respiratory bronchiole

Pleura

Alveolar duct—passageway from the bronchioles to the alveoli

How Lung Color Can Change with Age

Although lungs are pinkish in color during childhood, they can become grayish-pink later in life as a result of inhaling various environmental pollutants and other contaminants. Each day, your respiratory system is exposed to about 10,000 liters of air that contains chemicals, bacteria, viruses, dust, and other contaminants.

The figures to the right show the possible color changes in the lung from infancy, to middle age, and finally to old age. These changes would vary from person to person and depend on whether the person lived in the city or the country, what their occupation was, and if they were exposed to any other pollutants.

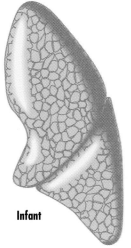

Infant

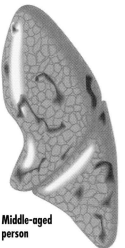

Middle-aged person

Elderly person

Asthma and Pneumonia

About 5% of all people have asthma at some point in their life. Asthma in children and young adults is generally triggered by plant pollen or dust, while asthma in older people is often provoked by environmental irritants, such as smog. In the early stages of this disease, the small bronchi are plugged with thick mucus. In the later stages, the bronchi and blood vessels are inflamed or thickened, and there can be fluid in the bronchial tissues. Asthma is an episodic disease, with each attack lasting minutes to hours. The disease can be mild or severe, and is sometimes even fatal. During an attack, there is a reduction in the expiratory flow rate and a decrease in the oxygen content of the blood.

Pneumonia can affect any portion of the lungs; it can be mild or severe; and it can be caused by infectious organisms (such as bacteria and viruses), by irritants (such as chemicals), or by food particles or large amounts of saliva that are accidentally aspirated (inhaled) into the lung rather than being prevented (by the epiglottis) from entering the lung. Common symptoms of pneumonia include fever, cough, and chest pain. The diagnosis is usually based on a physical examination and a chest x-ray, although other tests may be required.

Some Things That Can Cause Allergies and Asthma

This tiny organism, the dust mite (as seen under a microscope), is found in common house dust. In some people, it produces an allergic reaction, such as sneezing or an asthma attack.

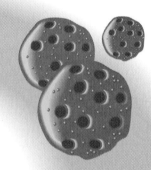

Pollen grains (shown magnified) are also potential triggers for an asthma attack or other allergic reaction.

People who are allergic to dogs or cats are usually sensitive to the dander (old skin cells that fall off), saliva, or urine of these animals. Immediate hypersensitivity can occur with the symptoms being either mild or severe. Direct contact with the animal is not required to produce the allergic response—only that the person is in an area where the dander, saliva, or urine has collected.

Environmental pollutants, such as concentrated car exhaust and smog from factory exhausts, can increase or precipitate allergic reactions and breathing difficulties. This is particularly true in people with asthma or lung diseases. However, if the pollutant is concentrated enough, it can also affect the general population.

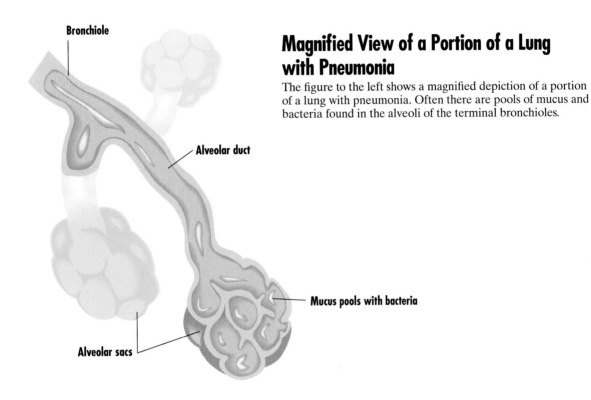

Bronchiole

Alveolar duct

Mucus pools with bacteria

Alveolar sacs

Magnified View of a Portion of a Lung with Pneumonia

The figure to the left shows a magnified depiction of a portion of a lung with pneumonia. Often there are pools of mucus and bacteria found in the alveoli of the terminal bronchioles.

Normal Lung Tissue vs. Lung Tissue with Pneumonia

Both figures below show examples of lung tissue as if viewed under a microscope. The lung tissue on the left contains normal cells (alveoli), while the lung tissue on the right shows abnormalities that are caused by pneumonia. These abnormalities consist of fluid and white blood cells inside the lung cells, as well as edema (fluid) collecting in the lung tissue. All or parts of the lungs may be affected when pneumonia occurs.

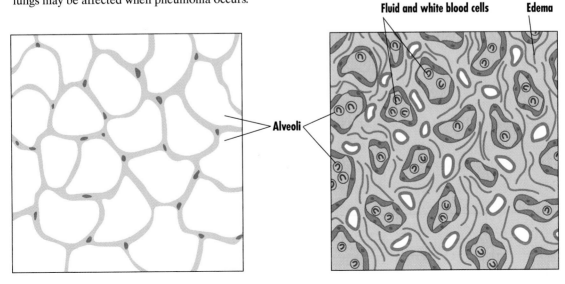

Fluid and white blood cells

Edema

Alveoli

The Lungs and Smoking

An estimated 140,000 people die of lung cancer each year in the United States, and about 85% of these deaths are attributed to smoking. More people die of lung cancer that is related to smoking than of any other type of cancer. Since there has been an increase in the number of women who smoke, the rate of lung cancer in women is increasing faster than it is in men. Lung cancer is now the leading cause of cancer-related deaths in women.

Smoking irritates the bronchi, can constrict the bronchioles, and causes some swelling of tissues. It also causes an increase in the secretion of mucus and other fluids, and since smoking slows down the movement of the cilia, this decreases the ability of the cilia and mucus to remove contaminants.

Emphysema, which is a lung disease, literally means an overabundance of air in the lungs. Its most common cause is cigarette smoking. In the United States, about 70,000 people die each year from emphysema and bronchitis, and an estimated 80% of these deaths are attributed to smoking. Early symptoms of emphysema may include cough, difficulty in breathing, and increased production of mucus and saliva. These symptoms worsen over time, and it becomes progressively more difficult to breathe. The respiratory rate increases, the vital capacity decreases, and the expiratory flow rate decreases.

A Normal Lung and a Lung of a Smoker

Lung tissue is normally elastic and dark pink in color, as shown in the tissue on the left. In contrast, the lung tissue in people who smoke is flabby and blackened, contains carbon particles, and has blocked airways, as shown in the drawing on the right.

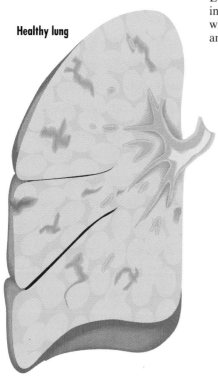

Healthy lung

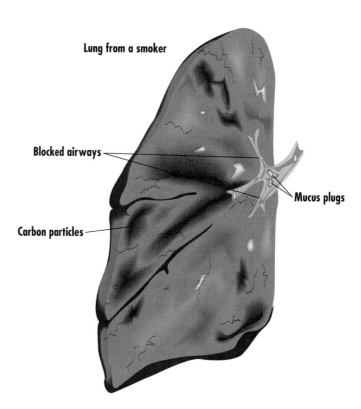

Lung from a smoker

Blocked airways

Mucus plugs

Carbon particles

Image courtesy of Lutheran General Hospital, Park Ridge, IL., and Dr. John P. Anastos, D.O.

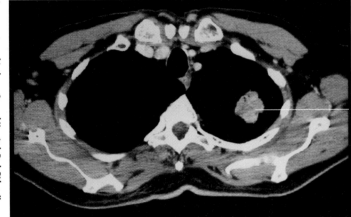

CT (Computerized Tomography) Scan of Lung Cancer

This CT scan shows a cancer in the left upper lobe of the lung.

Lung cancer

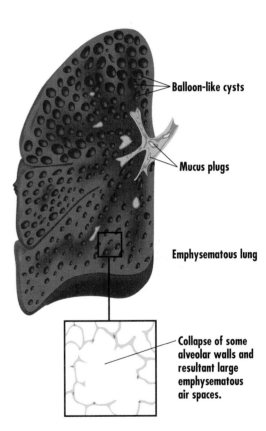

Balloon-like cysts

Mucus plugs

Emphysematous lung

Collapse of some alveolar walls and resultant large emphysematous air spaces.

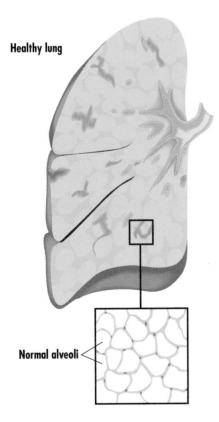

Healthy lung

Normal alveoli

A Normal Lung and a Lung with Emphysema

The drawing on the right shows how tissue from a lung affected by emphysema looks when viewed with the naked eye. The tissue is red-black and contains carbon fragments. Several large cysts, or balloonlike structures, are typically seen and are the result of destruction of the alveolar walls. In comparison, the normal lung tissue shown to the left is pinkish in color, smooth, and has no carbon particles.

While normal lung tissue appears compact, as shown in the microscopic view of normal lung tissue on the left, emphysematous lung tissue shows large air spaces, as shown in the example on the right. The over-inflated lung of a person with emphysema has lost its elasticity, the lung has difficulty contracting, air becomes trapped, and further air exchange becomes more difficult.

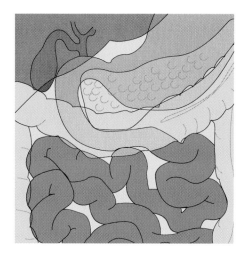

CHAPTER

12

The Stomach, Intestines, Gallbladder, and Liver

YOUR ABDOMEN AND pelvis contain many important organs critical for your survival. The abdomen consists primarily of the continuation of the digestive system organs from the esophagus in your chest, which then leads into the stomach, small intestines, and then the large intestines. The true abdomen also contains the liver, gallbladder, spleen, kidneys, pancreas, adrenal glands, and numerous blood and lymph vessels, lymph nodes, and nerves (the kidneys, pancreas, and adrenal glands being discussed, respectively, in Chapters 13 and 14). The true pelvis contains the remainder of the urinary tract (see Chapter 13), the reproductive organs (see Chapter 15), the last part of the colon and rectum, and blood and lymph vessels, lymph nodes, and nerves. These organs and tissues allow your body to (1) produce secretions that digest food, (2) move the food and utilize what it has digested, (3) store nutrients, (4) eliminate wastes and poisons through the urinary tract and the intestines, and (5) produce hormones, sexual characteristics, and the capability for reproduction.

Your digestive tract, or alimentary system, provides you with a continuous supply of water and nutrients. This system consists of the mouth, pharynx or throat, esophagus, stomach, and the small and large intestines. The liver and gallbladder are only indirectly involved, but they do secrete and store bile, which is involved in the digestive process. Digestion is accomplished through consuming food or drink by mouth; churning, mixing, breaking down, emulsifying, and absorbing the nutrients; and, finally, circulating the nutrients through the blood until they reach their final destination. It is at the site of these final-destination tissues or organs that the digested fats, proteins, carbohydrates, minerals, or vitamins are released, helping to keep your body healthy and working properly.

Chewing is essential for the digestion of solid food, since smaller food particles increase the speed at which the food can be further transported and digested. When your jaw muscles work to chew this food, the amount of force can be as much as 55 pounds in the front teeth and 200 pounds in the back teeth. The food then goes down your throat, into your esophagus (a muscular tube in your chest area), and to the stomach.

The stomach, located below the diaphragm, is a hollow muscular organ and has two openings—one entering from the esophagus and one going into the small intestines. At birth, the stomach's capacity is only about 30 ml, which increases to about 1500 ml in adults. The functions of the stomach are to temporarily store food, mix and help digest the food with stomach secretions and hydrochloric acid, produce an antibacterial action on its contents, and contract and squeeze the food down to the small intestines.

There is only a small amount of absorption from the stomach into the bloodstream, with most of the absorption occurring instead in the small intestines. However, water, alcohol, and sugar may be directly absorbed from the stomach, and this is one of the reasons that drinking alcohol on an empty stomach can produce an enhanced alcohol effect. Proteins, such as a T-bone steak, are partially broken down by secretions of the stomach. Not much work is required by the stomach for carbohydrate digestion, so this type of food proceeds quickly to the small intestines.

Most digestion takes place in the small intestines. The small intestines are divided into three parts. The first is the duodenum, followed by the jejunum and ileum, with their movements being a combination of mixing and propulsion. Both your small intestines and your mouth digest carbohydrates like rice and pasta. The small intestines also complete digestion of all the fats and proteins.

The appendix is a portion of the intestines where the small and large intestines join together. It is not believed to play a major role in the body, except when it becomes inflamed and appendicitis occurs. In cases of appendicitis, the appendix must be surgically removed to prevent massive body infection.

The next section of the digestive tract consists of the large intestines. They are made up of four sections—the ascending colon, which joins the small intestines, the transverse colon, the descending colon, and the sigmoid colon, which is s-shaped.

All necessary nutrients from the intestines (mainly the small intestines) are absorbed into the bloodstream, metabolized, and then transformed by the liver to be utilized elsewhere or stored there. The liver, the largest organ of the body, is reddish-brown and lies on the upper-right side of the abdomen. It is an important site for the storage of fats, carbohydrates, proteins, and vitamins. The liver is also crucial in that it forms and secretes bile, a mixture of bile acid and cholesterol, which emulsifies fats and neutralizes acid. Additionally, the liver absorbs poisons and toxic substances and neutralizes them.

The gallbladder is a pear-shaped, hollow organ closely attached to the back side of the liver. Its function is to store the bile secreted by the liver.

Some of the most common and talked-about abdominal illnesses, other than minor irritations or flu, include ulcers, gallbladder disease, colon polyps, and cancer.

Indigestion, while one of the most common health problems, is usually just a minor annoyance, and quickly dissipates. It usually occurs from eating too much or too quickly, or eating foods that irritate your stomach. The most common type of stomach disorder that usually does not go away without medical treatment is the ulcer. An ulcer is a scratch or deeper abrasion of the stomach or intestinal lining. There are two basic types of ulcers—stomach and duodenal (afflicting part of the small intestines). Duodenal ulcers are three to four times more common than stomach ulcers, occurring in about 6–15% of the population. However, the incidence of stomach ulcers is generally increasing, while the incidence of duodenal ulcers is decreasing.

Most medical experts now believe that stomach irritation or gastritis, probably ulcers, and some stomach and duodenal cancers may be the result of a bacterial organism called Helicobacter pylori. Also, factors such as genetics, emotions, stress, tobacco, alcohol, and certain drugs such as aspirin or others prescribed for arthritis may cause, precipitate, or aggravate ulcers.

The basic cause of inflammation or ulceration in the stomach or duodenum is not completely understood. However, people with duodenal ulcers are believed to have an imbalance between their protective barriers (of mucus production and secretions of the pancreas and intestines) and their secretion of highly acidic stomach juices. Therefore, these people may produce too much stomach juice or too little mucus. A somewhat controversial view is that stress or anxiety may also precipitate a duodenal ulcer. This may occur as a result of stress signals in the brain stimulating the vagus nerve (a cranial nerve—see Chapter 1), which in turn increases the acidic stomach secretions, which then empty into the duodenum and may produce an ulcer there.

In contrast, stomach ulcers usually do not occur in people with excess stomach secretions, but instead occur from the irritating effects of drugs, alcohol, or cigarettes or from a reduced resistance of the stomach lining to digestion.

Gallstones are a stonelike precipitation that may form in the gallbladder or in one of its passageways. These stones are usually very hard and can obstruct one of the passageways or the gallbladder itself. Approximately 20 million Americans develop gallstones, with about 35% of women and 20% of men having them. Additionally, gallstones have an especially high occurrence rate in Native American Indian women, of whom about 70% develop them. If the stone or stones remain in the gallbladder itself, then there may be few if any symptoms or problems. However, symptoms and problems usually do occur when the stones move and obstruct one of the ducts or passageways of the gallbladder. The symptoms can then include nausea, vomiting, pain, and fever. The most frequently used test to determine if there are gallstones is an abdominal ultrasound. The causes of gallstones are not completely known, however, there does seem to be a genetic predisposition.

Also, people with gallstones often have too much cholesterol in their bile (a product of the liver, employed in digestion and stored in the gallbladder). Other factors that may contribute to gallstone formation are a decreased amount of water in the bile and an inflammation of the lining of the gallbladder, as could occur from an infection. If the cholesterol and/or calcium and bilirubin can no longer remain in solution, small crystals can develop that may then coalesce to form a gallstone. The risk is also increased in women as they age, women taking estrogen or oral contraceptives, obese people, and people who have had a rapid weight loss.

Growths in the colon may give symptoms or signs of abdominal or rectal pain, bleeding from the rectum, changes in bowel habits, anemia, or fatigue. However, particularly with colon polyps, there may be no symptoms at all. Growths in the colon can be due to benign conditions such as polyps, or they may be the result of colon cancer. Colon polyps are growths or protrusions on the lining of the colon, and they may contain stalks. If their mechanisms of rapid cell reproduction become too disarranged, then cancer can form. Removing a polyp while it is still young usually prevents the area from turning into cancer.

Cancer of the colon can cause a variety of symptoms, some of which are listed above. Depending on its size and location, colon cancer can produce the following anatomic problems: ulceration, leading to bleeding in the stool and/or anemia; or obstruction, which can cause abdominal cramping, changes in stool caliber or size, constipation, or diarrhea.

One of the best ways to treat colon cancer is to prevent it. If a person is at risk for colon cancer, a doctor may do the following tests—a rectal exam, a test to check for rectal blood, a sigmoidoscopy using a lighted tube to view the sigmoid colon, or a colonoscopy, also using a lighted tube to examine the full colon. Many experts, including the American Cancer Society, recommend an annual rectal exam for all people over the age of 40 and a yearly fecal occult blood test with a flexible sigmoidoscopy every 3–5 years for all people at or over the age of 50. If there is a family history of colon cancer, then a colonoscopy or barium enema is recommended every five years beginning at the age of 35.

The following illustrations will present you with an inside view of your insides—the digestive system and its related organs. You will also see how alcoholism can affect your liver and other parts of your body.

The Digestive System and Related Organs

The Digestive System, Liver, and Gallbladder

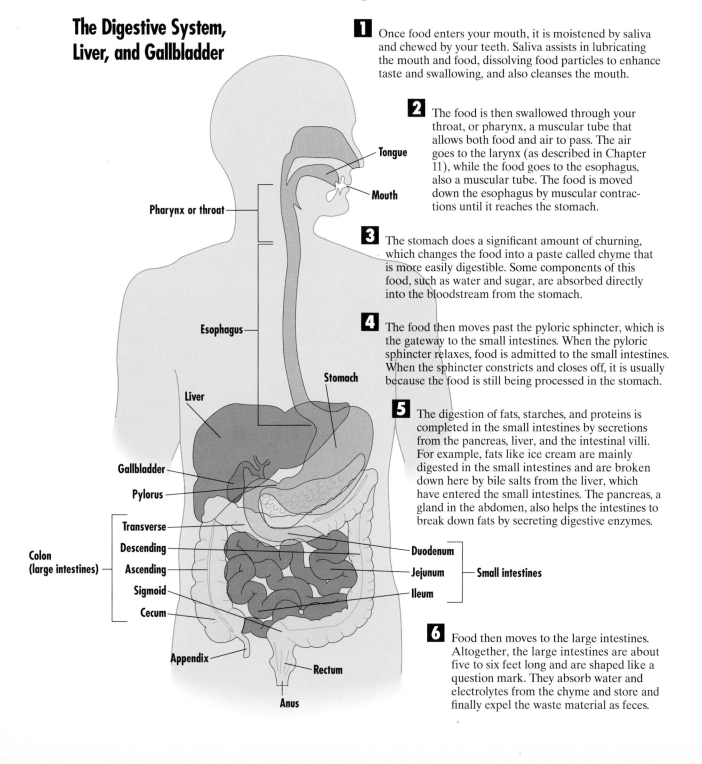

1 Once food enters your mouth, it is moistened by saliva and chewed by your teeth. Saliva assists in lubricating the mouth and food, dissolving food particles to enhance taste and swallowing, and also cleanses the mouth.

2 The food is then swallowed through your throat, or pharynx, a muscular tube that allows both food and air to pass. The air goes to the larynx (as described in Chapter 11), while the food goes to the esophagus, also a muscular tube. The food is moved down the esophagus by muscular contractions until it reaches the stomach.

3 The stomach does a significant amount of churning, which changes the food into a paste called chyme that is more easily digestible. Some components of this food, such as water and sugar, are absorbed directly into the bloodstream from the stomach.

4 The food then moves past the pyloric sphincter, which is the gateway to the small intestines. When the pyloric sphincter relaxes, food is admitted to the small intestines. When the sphincter constricts and closes off, it is usually because the food is still being processed in the stomach.

5 The digestion of fats, starches, and proteins is completed in the small intestines by secretions from the pancreas, liver, and the intestinal villi. For example, fats like ice cream are mainly digested in the small intestines and are broken down here by bile salts from the liver, which have entered the small intestines. The pancreas, a gland in the abdomen, also helps the intestines to break down fats by secreting digestive enzymes.

6 Food then moves to the large intestines. Altogether, the large intestines are about five to six feet long and are shaped like a question mark. They absorb water and electrolytes from the chyme and store and finally expel the waste material as feces.

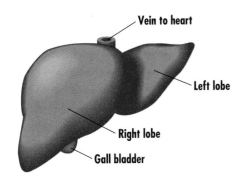

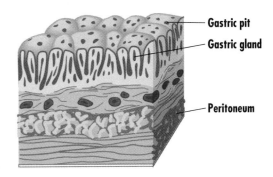

Liver

The liver is a reddish-brown organ, and it transforms and stores nutrients that have entered the bloodstream as part of the digestive process. The nutrients that the liver stores include fats, carbohydrates, proteins, and vitamins; it forms and secretes bile; and takes up and neutralizes toxins.

Stomach

The stomach consists of three layers of muscle covered by peritoneum, the outer membrane that surrounds abdominal organs like the stomach and lines the abdominal cavity. The inner surface of the stomach contains gastric pits that are composed of glands that secrete juices to help in digestion, and it also contains mucus, which protects the stomach lining.

Gallbladder

The gallbladder is a sac that lies beneath the liver and functions to store bile. Bile travels from the gallbladder to the cystic duct and then to the common bile duct, where it enters the duodenum of the small intestines to help in digesting fats.

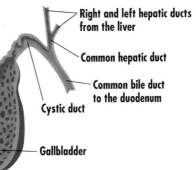

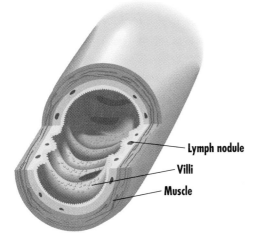

Small Intestines

The small intestines are about 22 feet long. They have circular folds, and their entire surfaces are covered by villi. These villi increase the amount of secretion and absorption that the small intestines can perform.

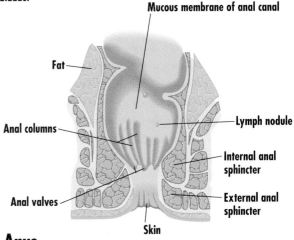

Anus

The anus is the opening of the lower part of the digestive tract from which waste material leaves the body. It is about 1½-inches in length. Hemorrhoids are a common disorder of this area and are due to bulging varicose veins on the inside or outside of the anus.

Gallstones, Stomach Ulcers, Colon Polyps, and Cancer

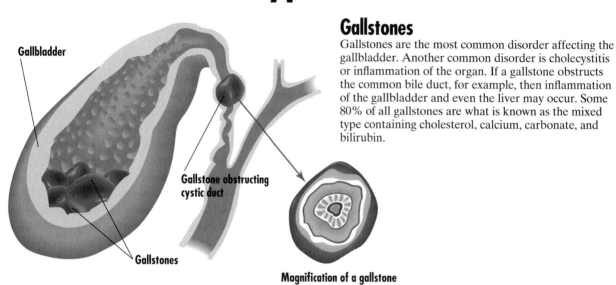

Gallbladder

Gallstone obstructing cystic duct

Gallstones

Magnification of a gallstone

Gallstones

Gallstones are the most common disorder affecting the gallbladder. Another common disorder is cholecystitis or inflammation of the organ. If a gallstone obstructs the common bile duct, for example, then inflammation of the gallbladder and even the liver may occur. Some 80% of all gallstones are what is known as the mixed type containing cholesterol, calcium, carbonate, and bilirubin.

Stomach Ulcer

Stomach ulcers, also known as gastric ulcers, are usually the result of either an injury (as from aspirin or arthritis drugs), or a defect in the stomach lining's resistance. The types of drugs prescribed for arthritis produce a large proportion of these kinds of ulcers because they reduce the stomach lining's defenses. In fact, stomach ulcers are increasing in frequency, while the other type of ulcers—duodenal or small intestinal ulcers—are decreasing in frequency, possibly due to an increase in the number of people taking arthritis drugs and aspirin, which can irritate or produce ulcers in the stomach.

Most people with stomach ulcers have pain, heartburn, indigestion, nausea, and/or a loss of appetite. The pain accompanying these types of ulcers may frequently increase after meals because the food stretches the organ, which then increases pain reception. Bleeding is also more frequent with stomach ulcers, and the blood travels down the intestines, where it can then show up as blood in the stool.

Emptying of the stomach contents can be delayed with stomach ulcers. This delay can cause regurgitation of the contents of the intestines and can produce injury and ulceration of the stomach tissues.

A stomach or duodenal ulcer can be diagnosed by a barium x-ray of the stomach or duodenum, or an endoscopy test, whereby a doctor looks into the stomach or duodenum with a lighted tube.

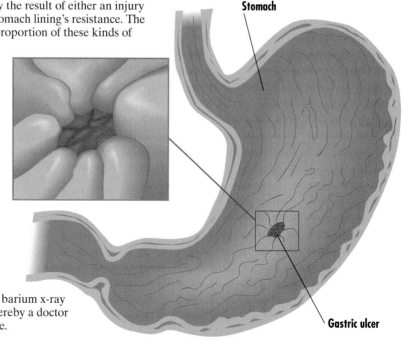

Stomach

Gastric ulcer

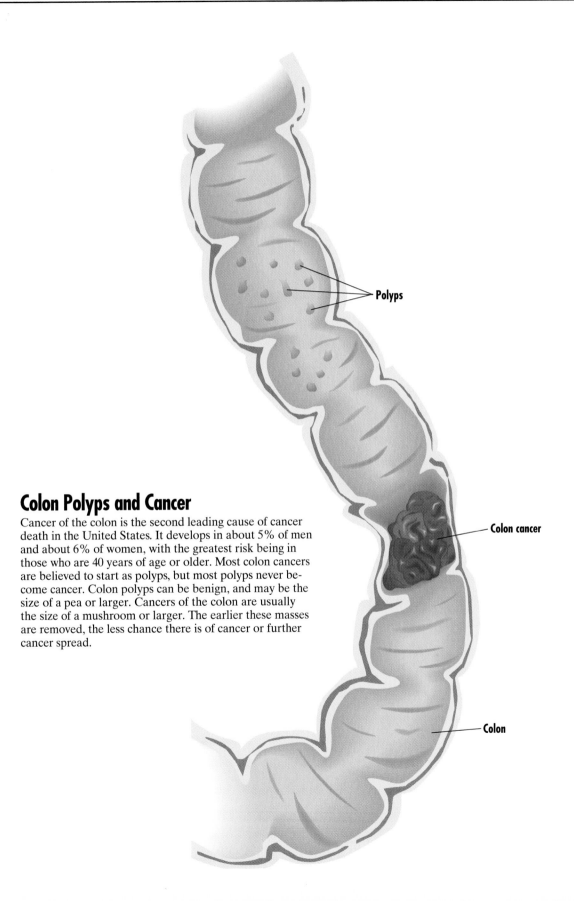

Polyps

Colon cancer

Colon

Colon Polyps and Cancer

Cancer of the colon is the second leading cause of cancer death in the United States. It develops in about 5% of men and about 6% of women, with the greatest risk being in those who are 40 years of age or older. Most colon cancers are believed to start as polyps, but most polyps never become cancer. Colon polyps can be benign, and may be the size of a pea or larger. Cancers of the colon are usually the size of a mushroom or larger. The earlier these masses are removed, the less chance there is of cancer or further cancer spread.

Alcohol and Its Possible Effects on the Body

Alcoholism is a chronic and potentially fatal disease. About 90% of Americans drink alcohol, with an estimated 40–50% of men having at least temporary alcohol-induced problems. Furthermore, 10% of men and 3–5% of women develop alcoholism.

Alcoholism can result in many health problems, which may occur even in the early stages of alcoholism. The severity and number of different symptoms and diseases increase with the length of time a person has been excessively drinking alcohol. The list below shows some of the potential health problems that can result from excessive alcohol drinking. Other problems can include anemia, decreased white blood cell count, heart rhythm irregularities, decreased size of the ovaries, infertility, spontaneous abortion, and fetal alcohol syndrome. Cancer is the second leading cause of death in alcoholics. Their rate of cancer is ten times higher than that expected in the general population, with the liver, pancreas, esophagus, stomach, or head and neck most often affected.

Potential Problems Resulting from Excessive Alcohol Consumption

Hallucinations, paranoia, memory loss, strokes, or brain damage

Yellow sclera of eyes

Painful, swollen muscles

Inflammation of the esophagus with bleeding

Enlarged spleen

Stomach inflammation and bleeding

Pancreas inflammation

Protuberant abdomen

Abnormal liver (cirrhosis, fatty liver, cancer)

Numbness, tingling of the legs and arms

Jaundice or yellow skin

Swollen legs

Alcohol and Liver Disease

Alcoholic liver disease is also a major health problem, with alcoholic cirrhosis a leading cause of illness and death in the United States. The liver's earliest response to alcohol is the accumulation of fat in the liver cells, followed by alcoholic hepatitis, and then liver cirrhosis. A larger number—90–100%—of heavy drinkers have fatty livers, with 10–35% developing alcoholic hepatitis, and 8–20% developing chronic cirrhosis of the liver. Hepatitis or cirrhosis symptoms can include liver enlargement, tenderness, pain, loss of appetite, nausea, and fluid (or ascites) in the abdomen producing a protuberance. These diseases can also produce enlargement of the spleen and inflammation and bleeding in the esophagus and stomach.

The prognosis for a person with alcoholic liver disease improves if the drinking of alcohol is curtailed, before irreparable damage occurs.

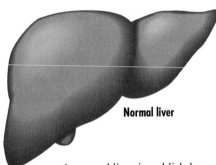

Normal liver

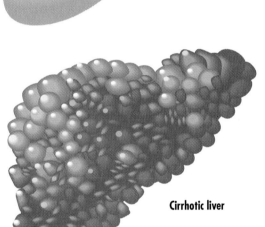

Fatty liver

A normal liver is reddish-brown, firm, and nongreasy. During the accumulation of fat in the liver cells from excessive alcohol intake, the liver increases in size and becomes soft, greasy, and yellow.

The word *cirrhosis* has its origins from the greek word "kirrhos" meaning "of yellow color." Cirrhosis is a diffuse, progressive process of the liver characterized by destruction of liver cells by fibrous tissue, followed by scarring and conversion of the normal liver tissue into abnormal hard nodules. The scar tissue impedes the flow of blood through the liver, increasing the blood pressure in the abdomen. This can then lead to multiple other problems including collection of fluid in the abdomen, enlargement of the spleen, dilation of the veins in the esophagus, swelling of the ankles, and kidney problems.

To the eye, the cirrhotic liver looks very nodular, fatty, yellow-orange, and scarred. Alternatively, the liver may instead be small, shrunken, and hard, or large and fatty.

Cirrhotic liver

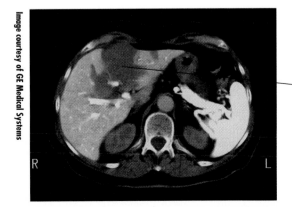

This figure shows a CT scan of a liver cancer. Liver cancer can develop from alcoholic liver cirrhosis

— Liver cancer

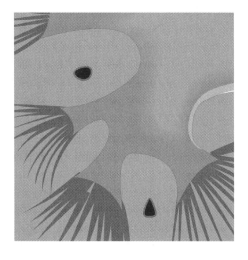

CHAPTER
13

The Kidneys and the Urinary Tract System

YOUR URINARY TRACT system consists of two kidneys, each with a ureter, which connects to the urinary bladder. The bladder then connects to the urethra, with the urethra ultimately leading to an opening out of the body. The function of this system is to produce and remove the waste product, urine, and to regulate the amount, alkalinity/acidity, and consistency of the body fluids.

The kidneys are oval-shaped organs, reddish-brown in color, about 4 inches by 2 inches wide, and lie in the upper part of your back. The kidneys are the blood's filters. Every minute, one-fourth of the body's blood—approximately 1200 ml—enters the kidneys. Each kidney is submerged in fat, which provides support. The adrenal glands lie on top of the kidneys (and they are further described in Chapter 14).

The kidneys are composed of about 2 million microscopic filters, called nephrons, which begin in the outer cortex, and continue into the inner medulla layers of the kidneys. These nephrons are responsible for absorbing nutrients and eliminating toxins and other waste materials from the blood.

The kidneys also help to regulate other bodily functions by secreting the hormones renin, erythropoietin, and prostaglandins. Renin helps to control blood pressure, and erythropoietin helps to stimulate the body to produce more red blood cells. In the kidneys, prostaglandins can cause dilation of the veins and are involved in the urine-making process. In other tissues and other situations, prostaglandins can cause smooth muscles to contract or relax. They can also be involved in abnormal fluid collection in the body, be responsible for some types of fevers and pain, and are heavily involved in the process of inflammation (such as with arthritis).

The concentrated urine from the kidneys moves into the ureters, and these tubes go to the bladder. The ureters and bladder are made up of smooth muscles, with the ureter being a transport tube, and the bladder also being a means of transport to rid the body of urine. Additionally, the bladder acts as a temporary storage area for the urine until it is ready to be released. At the base of the bladder is the urethra, which conducts urine out of the body. The male and female urethra differ in that a woman's urethra is short, while a man's is longer, traverses through the prostate, and conducts not only urine, but also sperm.

Urinary tract diseases can be caused by many problems. The most common problems include urinary tract stones, urinary tract infections, and incontinence. Other problems that can damage the kidneys are the harmful effects of diabetes, injury, cancer, certain drugs, or other illnesses.

Urinary tract stones result from a build-up of minerals or other chemicals that may occur at any level in the urinary tract, but usually first in the kidney. They may occur for no apparent reason, or they may be due to a specific problem. They usually become important only when they cause obstruction or bleeding, which may then produce symptoms of pain, decreased or absent urine flow, and blood in the urine. About 70% of the stones are calcium-containing, while 15% are "triple stones" made up of magnesium, ammonia, and phosphates. An additional 6% are uric acid stones, and 3% are cystine stones. Some experts believe that most calcium-containing stones are caused by heredity, glandular problems, dehydration, or a diet rich in oxalic acid. Foods rich in oxalic acid include spinach, coffee, and rhubarb. Other calcium stones may be the result of an illness of the parathyroid glands (see Chapter 14), or an infection. Uric acid stones are frequently caused by conditions that also cause gout. Gout is a condition that produces an extremely painful, hot, swollen joint or joints, and deposits of chalky crystals (called tophi) around the joint(s). Cystine stones result from a metabolic problem whereby the transport of this amino acid (a simple type of protein) is defective. In all, urinary tract stones are usually diagnosed by urine tests, x-rays, an ultrasound, or blood tests.

Urinary tract infections are very common in the United States and account annually for over 6 million health care visits. Infections can involve the ureters, bladder, or urethra, and they may be due to bacteria, fungi, or viruses. Bacteria are the most common cause of urinary tract infections and they are usually diagnosed by urine cultures. Women suffer from urinary tract infections more than men, usually because their urethra is shorter. The symptoms of urinary tract infections may include frequent urination, burning, fever, discomfort in the lower abdomen, and frequent urination at night. The likelihood of an infection increases during pregnancy, if urinary tract stones or tumors are present, or when men have prostate problems.

It is estimated that up to 20% of the elderly and up to 50% of nursing home residents have urinary incontinence, whereby they lose their urine without their voluntary control. Stress and urge incontinence are two common types of incontinence. Stress incontinence occurs most frequently in postmenopausal women and after pregnancy, but can also occur in men after prostatic surgery. In this type of incontinence, a small amount of urine is involuntarily released by actions such as coughing, laughing, or lifting an object. In urge incontinence, uncontrollable bladder contractions lead to a need to urinate, accompanied by an inability to hold the urine. About 70% of incontinence among older people is of this type. The causes include neurologic problems such as strokes, Parkinson's disease, or multiple sclerosis; urinary tract stones; urinary tract infections; or bladder or pelvic tumors. Urinary incontinence is important not only for the suffering it produces but also because it can lead to the breakdown of the skin, urinary tract infections, and broken bones from falls caused by slipping on the urine or having to go frequently to the bathroom.

Chronic kidney failure, a serious and potentially life-threatening disease, results from irreversible destruction of the nephrons and usually occurs over a period of years. It may be the result of diabetes, infections, other diseases, or certain medications. Over time, the inability of the body to rid itself of waste can lead to poisoning of the body, and without dialysis or a kidney transplant, the afflicted person can die.

The following illustrations will demonstrate the functioning of the urinary tract system, the flow of urine, and some common problems.

The Kidneys and the Urinary Tract

1 Urine is produced in the kidneys. Urine consists mainly of water but also salts, some body cells, waste products, urea, uric acid, and other nitrogen compounds produced mainly by the digestion of food proteins, foreign substances, and drugs.

2 Urine passes from the kidneys to the bladder through two ureters.

3 The urine empties into the urinary bladder, a temporary storage area.

4 When you decide to urinate, the brain tells the urethral muscles to relax and open. First the external sphincter then the internal sphincter of the urethra are told to relax, the bladder muscle walls contract, and the urine is then expelled.

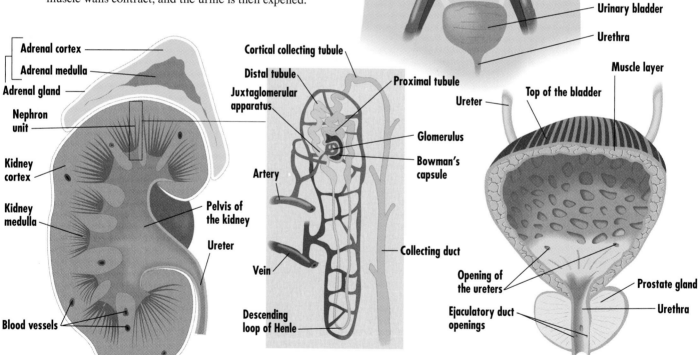

Labels (top right diagram): Vein to heart · Adrenal glands · Kidneys · Renal (kidney) arteries · Renal (kidney) veins · Abdominal aorta · Ureters · Urinary bladder · Urethra

Labels (kidney cross-section): Adrenal cortex · Adrenal medulla · Adrenal gland · Nephron unit · Kidney cortex · Kidney medulla · Pelvis of the kidney · Ureter · Blood vessels

Labels (nephron diagram): Cortical collecting tubule · Distal tubule · Juxtaglomerular apparatus · Proximal tubule · Glomerulus · Bowman's capsule · Artery · Collecting duct · Vein · Descending loop of Henle

Labels (bladder diagram): Muscle layer · Top of the bladder · Ureter · Opening of the ureters · Prostate gland · Urethra · Ejaculatory duct openings

This is a cross-section of the *kidney* and the *adrenal gland*. The kidney is divided into an outer cortex and an inner medulla, as is the adrenal gland. These two areas have different functions. It is in the cortex that blood enters the kidney, where it begins to be filtered and cleaned. By the time blood has reached the inner depths of the medulla, it has been completely filtered and concentrated and is now urine. The nephron is the main unit of function of the kidney.

The *nephron* of the kidney is composed of the glomerulus (a net of capillary arteries), the Bowman's capsule, and the tubules. At the beginning of the glomerulus lies the juxtaglomerular apparatus, which appears to be critical in regulating the flow and concentration of the urine. Each glomerulus is a filtering system from which filtered blood passes into different collecting tubules. In these tubules some substances are reabsorbed, other substances are secreted, and the concentration of what is now urine takes place. The urine is transferred to the collecting ducts, which then empty into the pelvis of the kidney and then the ureters.

The *bladder* stores urine until it is ready to be released. This shows the inside of a man's bladder. Most of the inner surface has irregular folds and ridges. The bladder is made up of three layers of muscles, providing the ability to expand and contract. Most people's bladders can hold about a pint of urine; when the bladder fills with urine, the walls expand and send impulses to the brain telling you to urinate.

Urinary Tract Infections and Kidney Stones

Infections of the urinary tract can be divided into those that occur in its upper part and those that occur in its lower part. Those infections that occur in the lower part usually set up housekeeping in the urethra or bladder or originate in the prostate. Those infections that occur in the upper part of the urinary tract predominantly involve infections of the kidney. If the infection reaches the kidneys (pyelonephritis), then abscesses may develop. If this has occurred, symptoms can escalate to more than just urinary frequency, pain on urination, or abdominal pain. There may be fever, chills, nausea, vomiting, diarrhea, muscle tenderness, and a fast heart rate.

Urinary tract stones may produce no symptoms, but usually they result in pain and bleeding with passage of the stone. Stones usually occur when the urine becomes supersaturated with minerals, when the person is dehydrated, or when excretion rates are excessive. Stones are more common in men, usually begin at the age of 30 or older, and may be genetic or due to a glandular problem.

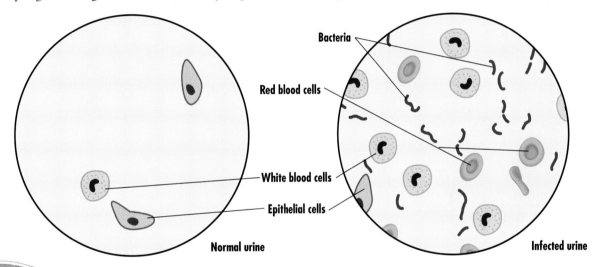

Bacteria

Red blood cells

White blood cells

Epithelial cells

Normal urine

Infected urine

The upper-left figure shows what normal urine would look like under the microscope. Normal urine has few or no white blood cells, no bacteria, and no red blood cells. When a urinary tract infection is present, a urinalysis often shows an increase in white blood cells, bacteria, and possibly red blood cells, as seen in the upper-right figure. Additionally, urine cultures are done and can indicate the type of organism and which antibiotics will help cure the infection. Many different organisms can infect the urinary tract, *Escherichia coli* being one of the most common.

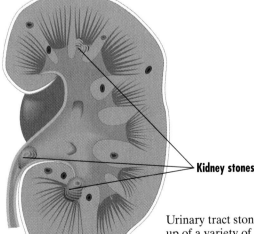

Kidney stones

Urinary tract stones, like the ones shown in this kidney, can be made up of a variety of different substances, including calcium or uric acid, or they can be a mixture called triple phosphate stones. Calcium is found in a large percentage of urinary tract stones.

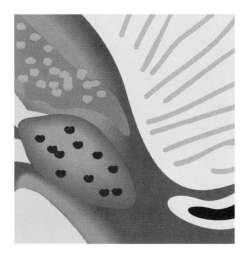

CHAPTER
14

The Thyroid, Pancreas, and Other Glands

YOUR BODY HAS two types of glands—exocrine glands that secrete fluids into a duct or a tube, and endocrine glands that secrete fluids directly into body organs. Some organs, like the pancreas, contain both endocrine and exocrine glands. It is because of these fluids, called hormones, that you can digest the tunafish salad you ate for lunch, that you have strong healthy bones that allow you to go for a long afternoon walk, that your body can break down the sugar in the Swiss chocolate candy bar with almonds that you devoured for a midafternoon snack, and that you can have sex and procreate a baby later that night.

Tissues containing exocrine glands, such as the skin, breasts, liver, prostate, stomach, or small intestines, are discussed in other chapters. In this chapter we will speak mainly about the endocrine glands. The endocrine glands include the pituitary, pineal, thyroid, parathyroids, thymus, pancreas, and adrenals.

Hormones, which are liquid chemical substances secreted by glands, are regulated by control mechanisms within the body. What happens is that the endocrine gland secretes its hormone, which then travels to receptors on cells that are within a particular tissue or organ. The tissue or organ can then carry out its function, but when the level of activity of the tissue or organ becomes too high, then there is a "negative feedback" to the gland, telling it to cut back on producing and secreting this hormone. An example of how this works can be shown with the pancreas gland, one of its hormones, insulin, and the target tissue, the blood, and ultimately muscle, fat, liver, and other body cells. After insulin is secreted from the pancreas, one of its actions is to prevent the level of sugar in your blood and body from becoming too high. Once this blood and body sugar level is lowered, a negative feedback tells the pancreas to stop secreting any more insulin until further notice. If the concentration of the insulin becomes too low, or is not perceived by the body, then a buildup of sugar can occur in your blood and body, and the result can be the disease diabetes mellitus.

Many experts consider the pituitary the main gland in the body. This reddish-gray gland is about the size of a pea (about 8 mm by 12 mm), is located at the base of the brain (see Chapter 1 illustrations), connected by a stalk to the hypothalamus. The pituitary increases in weight with pregnancy and decreases in old age. It has three main parts—the front, middle, and back. The front part, about 75% of the gland, secretes hormones that act to stimulate other glands and therefore these hormones and the pituitary gland itself have an indirect influence on most of the body. Also, this part of the gland is itself influenced by regulatory hormones released from the hypothalamus of the brain (see Chapter 1), which tell the pituitary when to release its hormones.

Six pituitary hormones are released from the front of the gland. Thyroid-stimulating hormone, TSH, stimulates the thyroid to produce its hormones; adrenocorticotrophin hormone, ACTH, stimulates the adrenal glands (on top of the kidneys) to produce hydrocortisone and other steroids; and growth hormone, GH, triggers the normal growth of all tissues and cells of the body, and in particular the long bones. Additionally, follicle stimulating hormone, FSH, causes the follicle cells of the ovaries to grow and also promotes the formation of sperm in the testes. Luteinizing hormone, LH, causes ovulation to take place in the ovaries and promotes the secretion of female sex hormones and testosterone by the testes. Prolactin induces development of the breasts and milk to be secreted from them.

The middle lobe of the pituitary gland may be involved in melanin secretion, which affects the pigmentation of your skin (see Chapter 5), and the back lobe of the pituitary produces two hormones—oxytocin and vasopressin. Oxytocin causes the pregnant uterus to contract, and vasopressin controls excretion of water.

Two diseases produced by abnormalities of the pituitary gland are gigantism and acromegaly. Gigantism is a condition found in children who have an excess production of growth hormone, and this disease can produce 8-foot giants. There is a proportionate enlargement of the bones, and the defect is often due to a tumor in the pituitary gland. If an excess production of growth hormone occurs in an adult, it is called acromegaly. This growth is disproportionate, with enlargement mainly affecting an increase in the size of the skull, prominent cheek bones, protruding jaw, and large, broad fingers and hands.

The pineal gland is also found in the brain. It's about 8 mm long, reddish-gray, and looks like a flattened cone. Its functions are still a bit mysterious, but it appears to modify the activities of the pituitary gland, pancreas, parathyroids, adrenals, and sex organs. It is also believed to produce two substances—melatonin and serotonin. Serotonin helps to promote a sense of well-being and restful sleep, while abnormal melatonin levels may contribute to depression, irritability, or lethargy.

The thyroid gland is brownish-red and consists of a right and left lobe, each about 5 cm long, connected by a narrow band of tissue called the isthmus. The thyroid gland varies in size and structure depending on factors such as sex, nutrition, age, iodine content, and temperature. It is one of the most sensitive organs in the body and produces three hormones—thyroxine, triiodothyroxine, and calcitonin. Thyroxine and triiodothyroxine are important for the metabolism functions of the body, and their purposes also extend to controlling the rate at which you absorb the fish you ate for lunch, speeding up your heart rate when necessary, increasing your respiration, increasing your mental activity, decreasing the quantity of cholesterol, and decreasing body weight. The thyroid hormones cause an increase in the activity of almost all the cells in the body, can increase the blood flow, and are believed to have a direct effect on the heart, which can then increase the heart rate. The quantity of these thyroid hormones that are produced is regulated by TSH, which is manufactured by the pituitary gland, with TSH being further controlled by the hypothalamus. Proper functioning of the thyroid is also dependent on the correct amount of iodine within the body, which the body must obtain from foods such as iodized salt or shellfish.

Most thyroid diseases can be divided into those that cause overactivation (hyperthyroidism) and those that cause underactivity (hypothyroidism) of the gland. In the illustrations you will learn more about the manifestations of two thyroid conditions—Grave's disease and hypothyroidism.

The parathyroid glands are small, yellow-brown structures lying on the back of the thyroid glands, measuring about 6 mm by 3 mm, and having a capsule that separates them from the thyroid glands. There are generally four parathyroid glands, but some people have as few as two or as many as six. These glands are important for secreting parathormone, which is crucial for the control of calcium and phosphorus.

The most common causes of a decrease in parathormone are the accidental removal of the parathyroid glands during neck surgery (although surgeons are careful to prevent it), hereditary causes, or an abnormal amount of magnesium in the body. The effects on the body from the decreased amount of parathormone include decreased blood calcium, which produces nervousness, spasms, and death unless treatment is provided.

Excess parathormone secretion usually results from a benign tumor, but can also be due to cancer. It causes an increase in calcium, with symptoms such as weakness, nausea, constipation, thirst, and kidney stones.

The thymus gland, which is smaller than a fist, lies behind the breastbone, and is normally large in children and shrinks with age. In young children, it is pinkish-gray in color, soft, and has lobules. After middle age, it becomes yellow because of an accumulation of fatty tissue. Its functions have not been clearly defined, but it is believed to be important in the regulation of your immune system—for example, in helping to produce lymphocytes, a type of blood cell (see Chapter 10) that helps to fight against infections and other immune diseases of the body.

The pancreas is a soft, grayish-pink gland, about 12 cm long, lying behind the stomach. The main part of the gland secretes enzymes that function in the digestion of foods and drinks. For example, the pancreas secretes trypsin and chymotrypsin which digest proteins, amylase which digests starches and other carbohydrates, and lipase which digests fats.

The pancreas also produces glucagon, which increases the amount of glucose (sugar) in the blood, and insulin, which decreases the amount of glucose in the blood. If there is a decrease in the amount of insulin produced, then diabetes mellitus results, which is the most significant disease of the pancreas.

The adrenals are two small, somewhat flat glands, each measuring about 50 mm by 30 mm and yellowish in color. Each adrenal lies above a kidney. The right adrenal gland is often triangular shaped, and the left one is often crescent shaped. The adrenal cortex, or outer part of the adrenal gland, is essential for life, and it produces hormones such as aldosterone, cortisol, and sex hormones. These hormones regulate body chemicals such as sodium and potassium; keep the blood at a proper concentration; allow the kidneys to work well; provide for sugar to circulate in the bloodstream; and allow the body to be able to cope with stress. If the concentrations of these hormones are too low, then the person can die unless proper treatment is begun. Most of these functions are due to the effects of the hormone aldosterone, particularly the regulation of fluid, potassium, sodium, and blood pressure. Cortisol increases the appetite, stablizes cells, is involved in the body's ability to handle stress, and in excess amount can promote collection of fat in the face, neck, and trunk. The most common cause of excess cortisol in the body is from the long-term use of prednisone or other steroids—for example, as a result of treating asthma or specific muscle or nerve diseases. It can also result from adrenal or pituitary tumors.

The adrenal medulla, or inner part of the adrenal gland, produces epinephrine and norepinephrine, both of which have similar functions. Epinephrine has the following effects on the body: It increases the heart rate and blood pressure, especially when stress arises. Additionally, it causes the flow of blood to the muscles to increase, the skin to become pale, the pupils of the eye to dilate, and smooth muscles to relax. Norepinephrine constricts the blood vessels, increases the heart activity (though less than epinephrine), slows down movement of the intestines, and decreases the blood flow to the intestines, liver, and kidneys.

The following pages will give you more information on the glands, diabetes, and thyroid disease.

The Endocrine Glands

The Pituitary Gland

The front part of the pituitary gland is important for producing hormones that stimulate the thyroid, adrenals, testes, and ovaries; encourage the growth of the body; and stimulate milk secretion by the breasts. The intermediate lobe may be involved with melanin secretion, which affects skin color. The back lobe of the pituitary produces a hormone that causes the pregnant uterus to contract, and another hormone that affects the excretion of water.

The Adrenal Glands

The adrenal glands are essential for functions such as the body's chemical regulation of sodium and potassium; blood concentration; pulse rate; smooth muscle relaxation or contraction; and dilation of the pupils of the eyes. The adrenal glands each comprise two different parts—the outer cortex, which is essential for the body's chemical regulation, and the medulla, or inner portion, which secretes epinephrine and norepinephrine, important in the "fight or flight" response.

The Pancreas

The pancreas is nestled next to the duodenum of the small intestines. It contains cells that secrete digestive enzymes and other cells that produce insulin and glucagon. Insulin is important for decreasing the amount of sugar in your blood, and glucagon is important for increasing the amount of sugar in your blood.

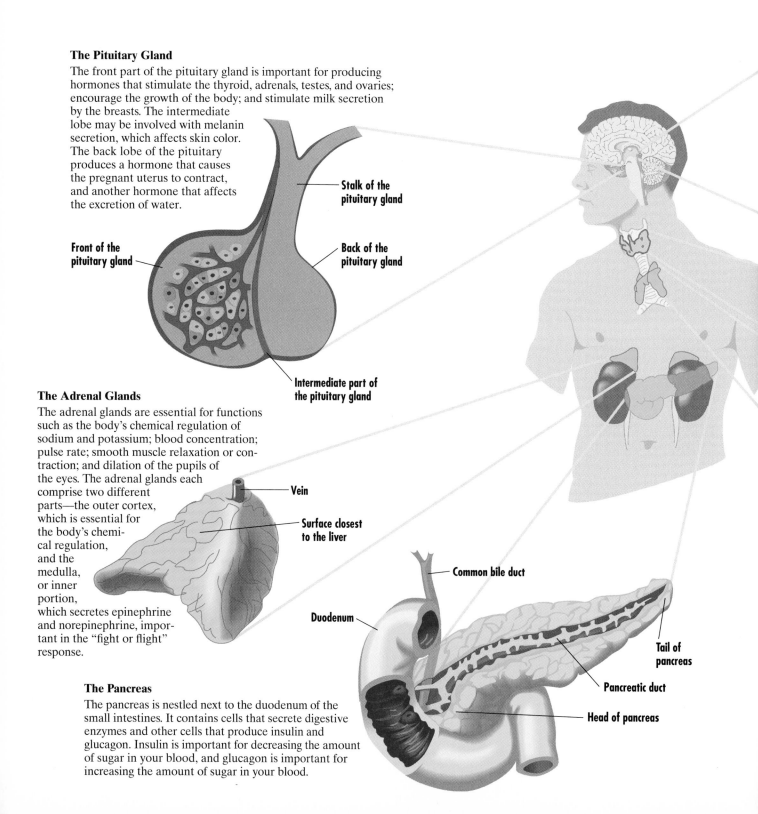

Stalk of the pituitary gland

Front of the pituitary gland

Back of the pituitary gland

Intermediate part of the pituitary gland

Vein

Surface closest to the liver

Common bile duct

Duodenum

Tail of pancreas

Pancreatic duct

Head of pancreas

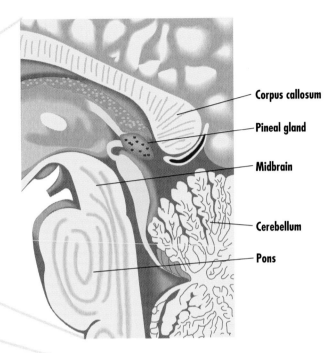

Corpus callosum

Pineal gland

Midbrain

Cerebellum

Pons

The Pineal Gland

The pineal gland may function in hormonal regulation, menstruation, and sex development. This gland contains sandlike granules whose functions are not yet known. The granules tend to increase with age.

The Thyroid Gland and Parathyroid Glands

The thyroid gland is divided into a right and left lobe and connected by an island of tissue called the isthmus. Frequently, the isthmus may have an additional, but still normal, tissue mass coming off of it—the pyramidal lobe. The thyroid produces hormones that function in the metabolism of the body and, to a lesser degree, the regulation of calcium in the blood.

There are usually four parathyroid glands, although there can be more or fewer. They primarily control the amount of calcium and phosphorus in the body and help with the development of bone. The main structural unit of the thyroid is the follicle, which consists of cells enclosing a cavity filled with a gel-like material. This gel is made up of proteins, enzymes, and thyroglobulin—a protein with iodine in it.

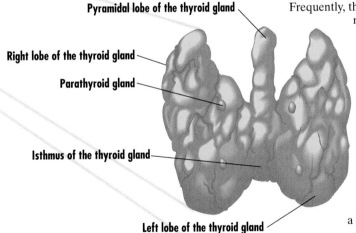

Pyramidal lobe of the thyroid gland

Right lobe of the thyroid gland

Parathyroid gland

Isthmus of the thyroid gland

Left lobe of the thyroid gland

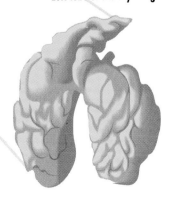

The Thymus Gland

The thymus gland functions in the immune system.

Diabetes Mellitus and Thyroid Problems

One of the most significant diseases of the pancreas and the body is *diabetes mellitus,* or simply *diabetes.* Some experts estimate that there are about 7 million people that have been diagnosed with diabetes in the United States, and perhaps as many as 5 million more who don't know they have it. The causes of diabetes are not known, but heredity is believed to play a role.

Diabetes exists in two major forms, Type I and Type II. Type I, in which no insulin is produced by the pancreas, is often called insulin-dependent or juvenile diabetes. It first occurs in younger people, usually less than 25 years of age. Type II diabetes, also called noninsulin-dependent or adult-type diabetes, usually occurs in people over 40 years of age. In this type, the pancreas produces some insulin, but the tissues do not respond appropriately to it. In both Type I and Type II diabetes, if the person eats or drinks anything with sugar in it, the sugar cannot be digested or digested well; it builds up in the body and damage can result. If diabetes is severe enough and not treated, it can result in death.

Diabetes can be diagnosed by testing for the amount of glucose, or sugar, in the blood. With appropriate treatment, many diabetics can continue to lead productive lives into old age. However, untreated or poorly treated diabetes can result in many problems, as seen in the figure to the far right.

Grave's disease, a condition of the thyroid gland, occurs in women more than men. It is especially common among people in their thirties and forties and may run in families. The exact cause is not known; however, the immune system is known to attack the thyroid, and antibodies are produced, which then increase thyroid growth and activity. Grave's disease and other diseases of *hyperthyroidism* (overactivity of the thyroid gland) can usually be definitively diagnosed with blood tests measuring the hormone TSH and the two direct thyroid hormones.

One of the common causes of *hypothyroidism,* or underactivity of the thyroid gland, is the after-effect of treatment for Grave's disease. It can also occur from other autoimmune problems, like rheumatoid arthritis or lupus. If this happens at birth, it is called cretinism; if not detected and treated, it can result in physical and mental deterioration. Both juvenile and adult hypothyroidism can be definitively diagnosed with a blood test for the TSH level.

Some common symptoms of diabetes mellitus and thyroid problems are listed in the figure to the right.

Common Symptoms of Diabetes Mellitus and Thyroid Problems

Diabetes Mellitus
Frequent urination
Persistent thirst
Increased appetite
Weakness

Grave's Disease
(hyperthyroidism—overactive thyroid)
Enlarged thyroid gland
Staring, bulging eyes
Swelling around the eyes
Nervousness
Heart palpitations
Muscle weakness
Excess perspiration
Hand tremor
Sleeplessness
Hair loss
Changes in the menstrual cycle

Hypothyroidism
(underactive thyroid)
Lethargy
Cold intolerance
Constipation
Slowing of the intellect
Slowing of muscle capabilities
Weight gain
Dry skin
Puffiness of the face
Thinning of the hair

Potential Problems Resulting from Untreated or Poorly Treated Diabetes Mellitus

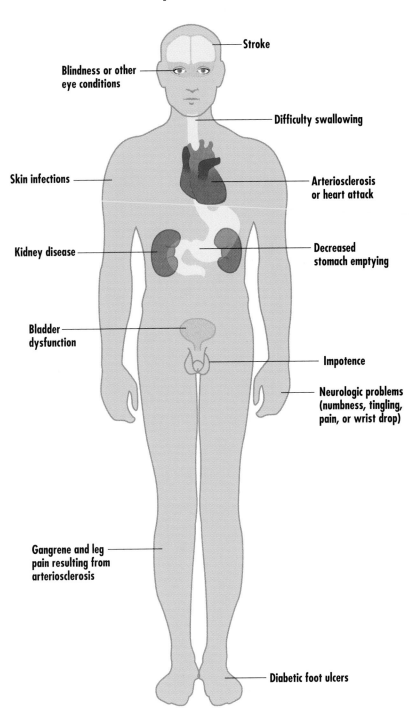

Stroke

Blindness or other eye conditions

Difficulty swallowing

Skin infections

Arteriosclerosis or heart attack

Kidney disease

Decreased stomach emptying

Bladder dysfunction

Impotence

Neurologic problems (numbness, tingling, pain, or wrist drop)

Gangrene and leg pain resulting from arteriosclerosis

Diabetic foot ulcers

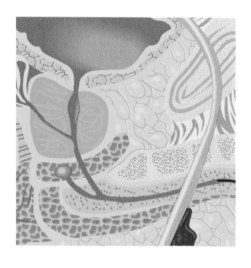

CHAPTER 15

The Male and Female Genitalia

THE MALE AND female bodies have very different genitalia with different hormones and different functions. Women menstruate and can form and harbor a new life, and men produce sperm, which are also essential in the formation of this new life. The main male sexual organs consist of a penis, two testes, the prostate gland, and a system that connects these and allows ejaculation of sperm out of the penis. The primary female sex organs are the external genitalia, vagina, uterus, two fallopian tubes, and two ovaries. The female breasts are considered to be a supplementary sex organ.

Sexual development begins in boys around the age of 12, at which time the hypothalamus of the brain tells the pituitary gland (also of the brain) to secrete two hormones, which in turn stimulate the testes to release additional hormones called androgens. Testosterone is the most important of these androgens. Its functions include sperm production, enlargement of the sex organs, and development of sexual characteristics like growth of hair under the arms and in the pubic area, deepening of the voice, and increased development of bones.

One of the most important sources of impulse in the male for initiating sexual intercourse is the glans penis, or the head of the penis. From the penis, nerves carry messages to other nerves in the sacral plexus of the spine, and then back to the penis, where the arteries become dilated and filled with blood, and an erection occurs. The degree of erection is proportional to the degree of stimulation. The same stimuli also cause the bulbourethral glands to secrete mucus, which lubricates the area and makes intercourse more comfortable. When the sexual stimuli become very intense, then other nerves cause sperm and fluid to be pushed forward and ejaculated out of the body through the urethra.

The penis is shaped like a finger and consists of the urethra and three cylinder-shaped structures that are capable of erection. During sexual stimulation, when the cylinder-shaped erectile tissues become engorged with blood, the penis becomes hard and erect. The blood is not allowed to drain out via the veins because the veins are closed off by the surrounding tissues. Once ejaculation has taken place, the veins open up, the blood can flow out, and the engorgement and hardness of the penis dissipate.

The testes are walnut-shaped glands that reside in a protective sac called the scrotum. The primary functions of the testes are to produce sperm and hormones called androgens, which are later changed into testosterone. Once the sperm are manufactured, they then travel into a tube called the epididymis, and then into another tube, the vas deferens, where they are stored. The testes of the young adult male form about 120 million sperm each day. During intercourse, fluids from the prostate and seminal vesicles, and sperm from the vas deferens are pumped into the upper parts of the urethra, where they are mixed, and then come out of the urethra of the penis during emission. During an average ejaculation there are about 200–600 million sperm released. The sperm are slender, moveable bodies with a hair-like tail that can help them move at a rate of about 1–4 mm per minute.

The prostate gland straddles the urethra and is the size of a chestnut, doughnut-shaped, and gray to red in color. It contains a thin, milky fluid, which consists of many substances including fats, calcium and other minerals, citric acid, a clotting enzyme and another type of enzyme called acid phosphatase, and a protein called prostate specific antigen (PSA). Levels of the latter two substances can become elevated in the blood of men with some prostate diseases.

Girls mature sexually about two years earlier than do boys. Here again, the hypothalamus stimulates the pituitary gland to secrete hormones, which this time act on the ovaries to produce the hormones estrogen and progesterone. Estrogen starts the enlargement of the sex organs and controls development of the female sex characteristics, which include breast growth, wide hips, and hair under the armpits and in the lower genitalia area. Additionally, estrogen is important for the proper cyclical functioning of the uterus and has a potent effect on the growth and health of the bones (see Chapter 6). The most important function of progesterone is to help promote the normal cyclical changes of the uterus. Progesterone also promotes breast development and the normal functioning of the fallopian tubes.

The ovaries are somewhat flat, oval bodies that lie on each side of the uterus. They are attached to the walls of the uterus and abdomen by ligaments. In function, the ovaries are analogous to the testes. Before sexual maturation, the ovaries are small and soft, whereas during child-bearing years they are smooth and firm, and during menopause they shrink. The ovaries' three main functions are to produce the main reproductive cell, the ova (or eggs), to manufacture specialized cells called follicles that surround the ova, and also to produce estrogen and progesterone. At birth there are about 1,000,000 ova, by puberty there are only about 400,000, and at menopause, there may be few or none remaining. One ovum matures about every 28 days, and at midcycle, hormones from the pituitary gland cause it to leave the ovary. The ovum then enters the fallopian tube, and if sperm are present, the sperm and ovum may join together to eventually form a fetus (see Chapter 16). If no sperm are present, then the ovum will travel down to the uterus. The uterine wall becomes engorged with blood, and dead tissue from the endometrial (innermost) layer of the uterus, other debris, and the ovum are shed during what is called menstruation.

The fallopian tubes are a pair of structures that connect to the ovaries and lead into the uterus, which is a thick-walled, pear-shaped organ. At the bottom of the uterus is the cervix, which has an opening into the vagina. The vagina is a hollow canal into which the penis fits during intercourse. The external genitalia include the clitoris and the vulvar area. The clitoris surrounds the opening of the vagina. The clitoris is composed of erectile tissue similar to that in the penis, and is controlled by nerves that travel from the area of the sacrum to the external genitalia, where they then dilate arteries, and subsequently increase the accumulation of blood in the area. Other nerves travel to glands in the area called Bartholin's glands, which create an immediate secretion of mucus. This is responsible for much of the lubrication during intercourse, although more is also provided from within the vagina.

The uterus undergoes continuous and repetitive cycles about every 28 days. The first phase after menstruation is heavily dominated by estrogen, and occurs from about day 4 through about day 15 of the cycle. This growth phase involves the production of new cells that form on top of the thin surface layer of the uterus. The second phase is dominated by progesterone and occurs from approximately days 16 through 28. During this time the uterine glands and blood vessels become highly coiled and tortuous, and the cells secrete a mucoid fluid that is rich in glycogen. On about the 27th or 28th day, the uterus enters a phase in which the blood supply constricts and no further gland secretions take place. The menstrual phase begins on day 1 of the menses and continues until about day 4.

The female breasts are modified sweat glands derived from the skin. Each breast is composed of 15 to 20 separate lobes arranged in segments similar to an orange. Breast tissue, in addition to being composed of lobes, also contains ducts, all of which are surrounded by fat and other tissues. The ducts terminate in the nipple and can bring milk to the surface when a baby is being nursed.

Menopause marks the normal end of a woman's reproductive cycle and typically occurs between the ages of 40 and 53, but it can occur earlier or later. It results from the ovaries secreting less estrogen, running out of eggs, and becoming less responsive to the hormone FSH. At first, the pituitary gland produces more FSH, trying to maintain the normal estrogen level, but in the end, it does not succeed. As a result, ovulation becomes infrequent, menstrual periods become irregular, and, finally, complete menopause occurs, with no further periods. Menopause often causes a reduction in the size of the breasts and the uterus, the walls of the vagina can become thin and dry, and there can be an increased risk of heart disease due to the lower estrogen levels. However, other than the cessation of menses and the anatomical changes described, many women do not have any other significant symptoms with menopause. Nevertheless, in some women the decrease in estrogen is accompanied by palpitations, depression, fatigue, sleeping difficulties, headaches, excessive sweating, or hot flashes.

Many health problems can occur that affect men's and women's genitalia. We will briefly discuss menstrual cramps, endometriosis, cervical cancer, and breast cancer in women, and benign prostatic hypertrophy, prostate cancer and impotence in men.

It is estimated that about 32 million women in the United States experience menstrual cramps each month. They occur before or during menstruation. The pain may spread to the hips, lower back, and inner thighs and may also be accompanied by other symptoms such as nausea, weight gain, fatigue, headaches, dizziness, and more. Menstrual cramps are usually caused by body chemicals called prostaglandins, and the amount released is usually related to the intensity of the cramps. It is believed that prostaglandins produce cramps by blocking some of the normal blood flow to the uterus, which causes the uterus to contract. The contraction is then perceived by the body as a cramping type of pain. Cramps are often a normal aspect of menstruation, but in some instances they are signs of a disorder such as fibroids, an infection, or endometriosis. You should always discuss any symptoms with your doctor.

Endometriosis is a condition that causes endometrial tissue (which normally lines the uterus) to develop outside of the uterus. Painful growths sometimes form on the outside of the uterus, ovaries, fallopian tubes, bladder, bowel, vagina, or other areas. The exact cause is unclear, but about 15% of all women develop endometriosis before menopause. The symptoms can include pain before and during menstruation, pain during intercourse, heavy or irregular vaginal bleeding, and infertility.

Rates of cancer of the cervix have dropped by about 70% since the initiation in 1943 of a simple test called the pap smear. The reason is that if precancerous lesions are found, they can be removed before they turn into actual cancer. Still, about 15,000 American women develop cervical cancer per year, and about 4,600 per year will die from it. The deaths are predominantly due to late detection, for if the cancer is detected early enough, the cure rate is about 100%. Also, interpreting pap smears can at times be puzzling, even to the most experienced technician. It has been estimated in some studies that 15–40% of pap smears fail to detect the true diagnosis because the sample was inadequate or the lab misread the smear. Some laboratories are now using a new screening system called Papnet, which uses computer imaging to inspect specimens. This technique may cut down on the error rate. The American Cancer Society recommends the following frequency for pap smears: All females who are sexually active or age 18 or older should have an annual pap smear. After a woman has had three or more consecutive negative pap smears, the doctor may recommend that the pap smear be performed less often. However, many doctors will still recommend an annual pap smear, even if the results continue to be normal. Pap smear screening, which only determines cancer or precancer in the cervix of the uterus, should continue throughout a woman's life. Last, not all abnormal pap smears mean cancer. Therefore, ask your doctor to explain the results to you carefully.

Most breast lesions are benign; they are not cancers, and do not form into cancers. When breast lesions are cancerous, the cure rate can be high if the cancer is detected early, but the disease can be devastating when not detected until it has reached an advanced form. In the United States, the lifetime risk of developing breast cancer is about 12%, and there is about a 3.5% risk of dying from it. It is important for all women to do monthly breast exams, beginning at the age of 20 and continuing throughout life. The American Cancer Society recommends an initial mammogram by the age of 40, a mammogram yearly or every two years from the age of 40 to 50, and a mammogram annually over the age of 50. If your family has a history of breast cancer, you may be at higher risk. You should speak to your doctor about having a mammogram done sooner and more frequently, which may or may not be necessary.

Benign prostatic hypertrophy, or enlargement of the prostate, is one of the most common conditions affecting middle-aged and older men. By the age of 80, almost all men have this hypertrophy, although in some cases it may be small, cause no symptoms, and need no treatment. (See the illustrations for further information.)

Prostate cancer is now the most common form of cancer in men, with about 100,000 men newly diagnosed with this cancer each year. Fortunately, most men who get prostate cancer do not die from it, with only about 3% of American men being expected to succumb to it. (See the illustrations for further information.)

It is estimated that approximately 30 million men in the United States suffer from impotence, but the numbers may actually be higher than is reported because some men are still unwilling to talk about this condition with their doctors. In a recent study, 52%

of men between the ages of 40 to 70 said they suffered from chronic bouts of impotence. The incidence of reported impotence has increased over the past 10 to 15 years. This increase may in part be due to the aging of the population—older men are more likely than younger men to experience impotence. The increase in reported cases could also be the result of more awareness about health issues in general and knowledge that more has been learned about the causes of impotence with better treatment options available. Problems of the pelvic blood vessels are the most frequent cause of impotence, with neurologic (related to nerves in the area), structural, and psychological problems being responsible for a large proportion of the causes, and hormonal abnormalities being less likely.

If you turn the page to the illustrations, you will learn more of how men's and women's sex organs work, and some common related disorders.

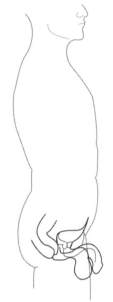

Men's Genitals

The penis is composed of three cylindrical bodies of cavernous erectile tissue. It is during erections that these cavernous spaces become engorged with blood, causing the penis to become hard and erect. The urethra, which begins in the bladder and carries the urine and sperm, goes through the penis. Other parts of the male genitals include the vas deferens, which is the continuation of the epididymis that began in the testes, and which stores the sperm. There are two seminal vesicles located between the bladder and the rectum. The prostate, a red-gray dense gland, produces other secretions that help form the ejaculate and may also ease the act of intercourse.

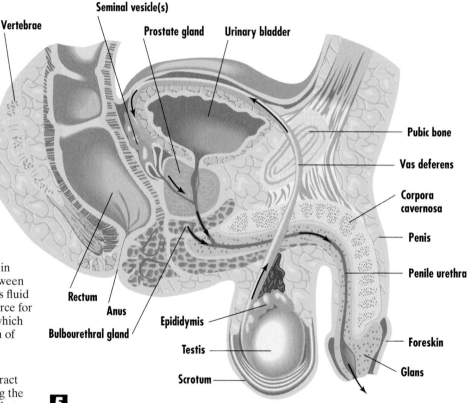

Seminal vesicle(s)
Vertebrae
Prostate gland   **Urinary bladder**
Pubic bone
Vas deferens
Corpora cavernosa
Penis
Penile urethra
Rectum
Anus
Bulbourethral gland
Epididymis
Testis
Foreskin
Glans
Scrotum

1 Sperm are formed in the testes and then travel to the epididymis. After being in the epididymis for about 24 hours, they are capable of movement.

2 Most sperm move from the epididymis to the vas deferens, where they are stored and remain viable for about one month.

3 Seminal fluid (semen) is formed in two seminal vesicles located between the bladder and the rectum. This fluid contains fructose, an energy source for the sperm, and prostaglandins, which are important in the fertilization of the egg by the sperm.

4 The seminal vesicles contract during ejaculation, forcing the seminal fluid into a duct that leads into the urethra.

5 Other secretions from the bulbourethral glands are added to the seminal fluid.

6 During ejaculation, the sperm travel from the vas deferens into the urethra, where they are mixed with the seminal fluid. The fluid is then released through the head of the penis.

Sperm

The head of a sperm contains most of the genetic material, the body helps to provide energy, and the tail is useful for movement. Most sperm travel in a straight line at about 1–4mm per minute. Although they can live about one month in a man, their life span in women is only about one to two days.

Tail

Body   **Neck**   **Head**

Women's Genitals

Each part of a woman's genitalia has a specific function. These include manufacture and maturation of eggs (ova) for reproduction, production of hormones that affect menstruation and menopause, preparing and aiding the body for possible pregnancy, and nourishment of newborn babies.

1 During their normal reproductive years, women undergo a monthly rhythmic cycle (on average 28 days) involving their female hormones and sexual organs. The first day of the cycle begins on the first day of menstruation (bleeding). Just after menstruation has begun, the levels of the FSH and LH hormones increase, allowing the maturation of immature egg cells in the *ovary*. Estrogen levels then increase, and after about seven days of growth, one egg predominates.

2 Approximately 14 days into the cycle, the egg is released from the ovary (in a process called ovulation) and goes into the *fallopian tube*. There are two fallopian tubes, one for each ovary, and each one leads into the uterus.

3 If fertilization by a sperm does not occur in the fallopian tubes, then the egg simply travels down into the *uterus* and waits to be discarded. The uterus is a hollow, muscular organ that lies between the bladder and the rectum. If the egg has been fertilized by a sperm, it will implant in the uterus and grow there into an embryo and then a fetus. If fertilization has not occurred, the egg will degenerate in the uterus.

4 Approximately two days before menstruation, estrogen and progesterone levels decrease, and menstruation then begins. The flow travels past the mouth of the uterus into the *cervix*.

5 The material, including the degenerated egg, goes into the *vagina*, a muscular tube, then leaves the vagina and the body, passing the *clitoris*, which is female erectile tissue located on the outside of the body.

Fallopian tube
Ovary
Uterus
Vertebrae
Cervix
Urinary bladder
Vagina
Rectum
Clitoris Urethra Anus

Pectoralis muscle
Intercostal muscle
Rib
Fat
Skin
Lobules within the lobes
Glands
Ducts
Nipple

Nucleus of egg
Egg
Follicle cavity

This is a microscopic view of a developing egg and its nucleus. Follicle cells form a cavity in the follicle, and layers of follicle cells congregate around the egg itself. Usually only one follicle and egg mature during each menstrual cycle.

The breasts are composed of glandular tissue, lobular tissue connected by fibrous tissue, and fatty tissue. The smallest lobules open into ducts, which serve as reservoirs for milk.

Breast and Cervical Cancer

Though the majority of breast lesions are benign, *breast cancer* is a common cancer of women. Definitive causes have not been identified, but some risk factors may include genetics, early age of menstruation, no biological children, late age for the first pregnancy, pesticides, and alcohol use. Yet, some studies have estimated that 70–80% of all breast cancers occur in women who have no risk factors, so clearly the primary causes may not have been isolated yet.

One of the best ways to detect breast cancer early is to do a monthly breast exam beginning at the age of 20 and continuing throughout life. Mammograms are also important in detecting breast cancer before it may even be noticeable on exam.

The causes of *cancer of the uterine cervix* are not known, but some risk factors include smoking, being a black or Hispanic-American female, or being HIV positive. Risk may also be related to the age of first sexual activity, the number of sexual partners, or being of a low socioeconomic status. Early symptoms can include bleeding, which may occur after intercourse, exertion, or bowel movements, or at any time.

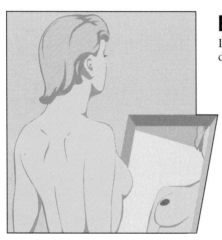

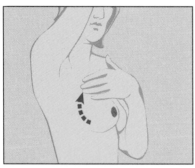

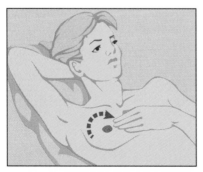

Breast Self-Exam

It is important that women do monthly breast self-examinations to detect any abnormal masses, thicknesses, lumps, or other changes.

1 Stand in front of a mirror and observe if there are any changes in the breasts such as scaling, discharge, or puckering.

2 Place your hands behind the head and press. Again, look in the mirror to notice any changes, as in step 1.

3 Bend forward with your hands on the hips and pull the shoulders and elbows forward. Again look in the mirror to notice any changes, as in step 1.

4 With one arm raised, move the opposite fingers on top of the breast (where the arm is raised). Beginning at the outer edge of the breast, move the fingers in a circular motion slowly around the entire breast until the nipple is reached, being careful to also include the armpit. Any abnormal masses or lumps should be noted and discussed with your doctor.

5 Gently squeeze the nipple to see if there is any discharge.

6 Then lying down, with the same arm extended over the head, and a pillow under that shoulder, re-examine the breast, as in steps 4 and 5. Steps 4 to 6 should then be repeated on the opposite breast.

Additionally, women should have a physician examine their breasts. The American Cancer Society offers the following guidelines for frequency of breast exam by a physician: every 3 years between the ages of 20 and 40, and yearly after age 40.

Enlarged Prostate and Prostate Cancer

Benign prostatic hypertrophy (BPH), an enlargement of the prostate gland, is the most common benign tumor found in men. The prostate is enlarged, and may have numerous nodules on it. The urethra narrows to a small slit and creates pressure on the urethra. Symptoms can include difficulty in starting the urine stream, decreased force and caliber of the stream, dribbling, sense of fullness after urination, frequency, urgency, pain on urination, and frequent urination during the night. In severe cases, as in the figure below, there can also be thickening of the bladder wall, dilation of the ureters, and a resultant enlargement of the kidneys.

Most *prostate cancers* are usually slow growing, although some can be aggressive. Both early and advanced prostate cancers may have no symptoms. If there are symptoms, men usually have urinary tract problems, such as frequency, burning on urination, and blood in the urine, or have back or hip pain. The causes are not known, but some potential risk factors may be heredity, a high-fat diet, and being an African-American man.

Signs of prostate cancer can often be detected by your doctor during an annual rectal/prostate exam. With prompt treatment, the survival rate in men with a small localized prostate cancer is about the same as for men who never had prostate cancer.

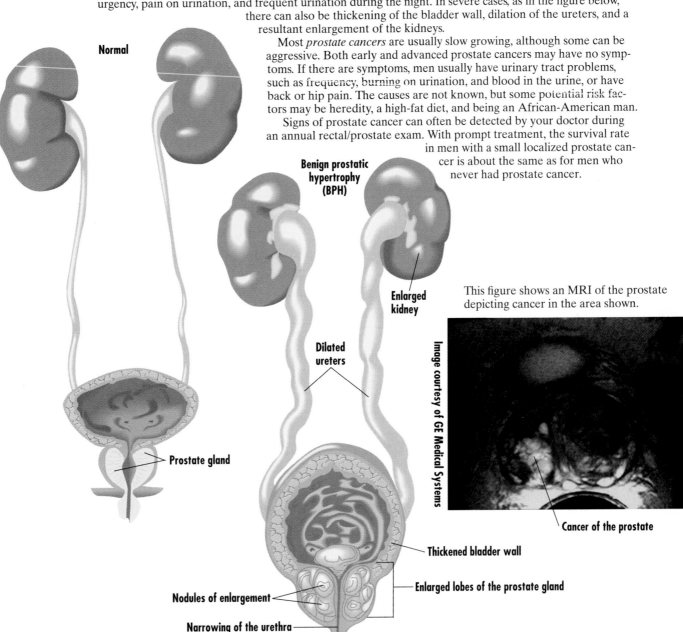

Normal

Benign prostatic hypertrophy (BPH)

Enlarged kidney

Dilated ureters

Prostate gland

This figure shows an MRI of the prostate depicting cancer in the area shown.

Image courtesy of GE Medical Systems

Cancer of the prostate

Thickened bladder wall

Enlarged lobes of the prostate gland

Nodules of enlargement

Narrowing of the urethra

Cells, Pregnancy, and the Formation of a New Life

THE WONDER OF a new life continues to amaze even the most jaded person. Many scientists say the more they learn about this subject, the more they marvel. At the moment you were first created, you consisted of a single cell. As time went on, you grew to consist of millions of cells, as you still do. Your cells die, and new ones are continuously created. Cells are the basis of your body—the smallest structural unit. They're typically separated into two parts: the cytoplasm (everything between the cell membrane and the membrane of the nucleus) and the nucleoplasm (everything within the nucleus).

The basic parts of the cell consist of the following: (1) the outer cell membrane, which allows or prevents materials to come in or out; (2) mitochondria, enzyme- and protein-rich sacs that produce energy through cellular respiration and that are found in all cells except mature red blood cells; (3) Golgi apparatus, cellular components that are important in producing secretions and concentrating them; (4) glycogen and lipids, used for energy and stored in sacs; and (5) one or more nuclei, each of which is supplied with DNA and RNA. Both RNA and DNA are complex, with a number of proteins and other components, and are important for the genetics of your body, determining that your eyes are blue, that your hair is black, and many other things about you. DNA (deoxyribonucleic acid) is the major source of genetic information in each cell of your body. It is estimated that your body has approximately 3 billion base pairs of DNA, and that each cell has 23 pairs of chromosomes, in which the DNA is found. Everyone inherits one copy of the genetic package from each parent. This determines your sex, ethnicity, and many other characteristics and influences your health. RNA (ribonucleic acid) is found in all cells and helps in the synthesis of proteins and the transmission of the genetic information found in the DNA.

Although the simplest animals consist of a single cell, other higher animals—including people—are a multiplicity of independent cells specialized and united to form four basic tissues—epithelial tissues (skin), connective tissues (such as tendons and bones), muscles, and nerves.

It is almost a miracle how these cells and tissues combine to form a new human being. If conditions are right, the road to the development of a new life can begin shortly after sexual intercourse, or more specifically, after a sperm fertilizes an egg. For fertilization to occur, some sperm must make it up to the fallopian tubes of the woman (see Chapter 15). They may be aided by contractions of the uterus and the fallopian tubes. One sperm must then penetrate through the layers of one egg. The now-fertilized egg then starts moving toward the uterus. This journey can take from three to four days. The egg stays in the uterus for two or more days until it has become a complex mass of cells (known as a blastocyst). On the seventh day, it implants itself into the uterine wall, where it will remain until it is delivered as a baby—about 265 days after ovulation and conception.

Throughout this time, development is constant. By the 16th to 18th day, the head begins to grow, and blood begins to flow inside the blastocyst groupings of cells. Some of these blastocyst cells form into projections, which become the placenta.

The placenta consists of villi, which are fingerlike projections into which the fetal blood vessels—capillaries—grow. Therefore, these villi contain blood from the fetus, but in addition are also surrounded by sinuses containing blood from the mother. The area of the placenta also contains two umbilical arteries (oxygen-poor blood) and one umbilical vein (oxygen-rich blood); all three blood vessels having run through the umbilical cord before entering the placenta. Oxygen and food, from the mother, diffuse directly into the villi and capillaries of the fetus. Waste materials and carbon dioxide from the fetus diffuse directly from the villi into the mother's blood, where they are then transported away to be eliminated. Therefore, the major functions of the placenta are to provide nutrition for the fetus from the mother's blood into the fetus' blood, and excretion of waste products from the fetus back into the mother for her to excrete. The placenta also serves as an endocrine organ and produces at least four hormones, including estrogen, progesterone, and HCG. HCG is the first hormone to appear, and it is actually the level of HCG in the blood or urine that is measured by a pregnancy test. In addition, the levels of progesterone may be four or five times higher in a pregnant woman's body. It is progesterone that allows the sperm to move up into the fallopian tubes more quickly, stimulates the maturation of the fertilized egg, and prevents the uterus from moving too much and expelling the fetus.

By the 22nd to 26th days, the head has already begun developing, the lower part of the abdomen has been formed, rudimentary arms and legs begin to grow, and the heart begins to develop. This is the beginning of what is called the embryo period. At the fifth week, the arms and legs have elongated, the heart has started beating, and the ears, hands, and feet have begun to form. By the sixth week, the liver begins to develop, and by the seventh to eighth week, the fetal stage has begun. It is during this time that the face begins to develop (with completion of the upper lip and nostrils), the external ears start to mature, and the external genitals begin to differentiate.

At the third month, the neck becomes longer, the eyelids meet and fuse, and nails appear. The four- and five-month-old fetus develops sebaceous glands, and a protective cheesy covering, the vernix caseosa, forms over the skin. At this stage, the sex of the fetus can be easily identified. Also, primary hair, or lanugo, develops, and the head and the upper arms are still disproportionately large compared to the rest of the body. The six-month-old fetus has wrinkled skin and well-developed eyelids and eyebrows. The seventh-month-old fetus has a body that is now more plump and round, the skin is less wrinkled, and the hair of the scalp is longer. In the eighth month, the body shape becomes more infantlike, and in the ninth month the lanugo hair usually disappears, although it may still be present at birth. Also, the umbilicus is almost in the middle of the body, the wrinkling is diminished, the body is more rotund, and the chest is prominent. By now, the baby is ready to come out. Look out world, here comes a brand-new person!

Development of a fetus also means that the mother's body is going through some tremendous changes. Many times the earliest obvious sign that a woman is pregnant is a missed menstrual period. Additionally, the woman may experience heaviness in the breasts, nausea, and increased frequency of urination.

A pregnancy test can be performed after the normal menstrual period has been overdue for one or more days. If this initial pregnancy test is negative, it should be repeated in seven to fourteen days, as these tests are more accurate at this later time. Also, a vaginal examination by the doctor can reveal an enlargement of the uterus. An ultrasound test, which is believed to be harmless to the fetus or the mother, can also define the pregnancy and can provide a radiological picture two weeks after a missed period. Later in the pregnancy, an ultrasound can also help detect multiple births or certain fetal disorders. If something looks suspicious, or if the mother is thought to be at particular risk, an amniocentesis may be performed. In this test, a sample of fetal (amniotic) fluid is extracted by inserting a needle into the mother's abdomen. An amniocentesis test may help reveal various genetic abnormalities.

It is important for a woman to see a health-care provider as soon as she thinks she may be pregnant. She will receive a complete physical exam, including blood pressure and urine tests. At each subsequent visit the doctor or nurse will usually check the blood pressure, do a urine test, check the ankles for swelling, and check the growth of the uterus. At first the visits are usually once a month, then from the 28th week they are usually scheduled every two weeks, and weekly from the 36th week on, depending on the woman's own situation and health conditions.

Different mothers react to pregnancy in different ways. Probably the most common complaint during pregnancy is "morning sickness," or nausea and vomiting. Although the term implies that the symptoms occur only in the morning, this is not necessarily true. These symptoms are most common during the first half of pregnancy. The cause of pregnancy-induced nausea or vomiting is not clear, but it is believed to be due to hormonal changes. Stress or emotional factors may also contribute. Usually the symptoms can be lessened by eating small amounts of food at more frequent intervals and avoiding foods that seem to aggravate the symptoms.

All pregnant women develop an enlarged uterus to hold the fetus, enlarged breasts, and an enlarged vagina. The average weight gain is about 15 to 24 pounds, with most of the gain occurring in the last two-thirds of the pregnancy. Also, changes in hormone levels can cause the legs to swell and can produce the temporary development of acne and some masculine features.

Other problems that commonly occur, especially in the latter half of pregnancy, include the following: backaches, headaches, constipation, frequent urination, heartburn, ankle swelling, hemorrhoids, sleeplessness, palpitations, and sweating. Pregnant women often have backaches because the ligaments that normally hold the joints in place are more stretched and relaxed as an effect of the hormones released during pregnancy. Headaches may result from fatigue, stress, or the change in hormones. Constipation can be produced by the increased production of progesterone. Frequent urination may result from increased pressure on the bladder, or it may be due to infection, which is more likely to occur during pregnancy. Progesterone causes the relaxation of muscles at the lower end of the esophagus, which allows stomach acid to reflux back into the esophagus, causing heartburn. Ankle swelling is produced by the effects of progesterone on the blood vessels and the pressure and weight of the uterus on the veins that carry blood from the legs, which can also cause varicose veins. Hemorrhoids usually result from constipation. Sleeplessness often occurs during the last few weeks of pregnancy and can be due to the discomfort caused by the enlarged abdomen, backaches, or fetal movements. Palpitations and sweating may result from the imbalance of normal hormone levels.

The onset of labor begins with strong regular contractions every 20 to 30 minutes, with a dull ache in the lower abdomen and back. There may be a discharge of a small amount of blood and mucus as the cervix starts to open. There also may be a normal rupture of the fetal membrane, with a rush of fluid from the vagina.

Labor has three stages. In the first stage, the frequency of the contractions continues to increase until the cervix is fully open, which commonly lasts from five to ten hours, particularly if it is the mother's first child. The pain experienced by the mother during this stage is probably caused by a lack of oxygen to the uterine muscles, resulting from compression of the blood vessels to the uterus. In the second stage of labor, usually lasting up to about two hours, the contractions will occur every one to two minutes. This stage can be physically and emotionally strenuous. The baby is coming through the birth canal, and the stretching of the cervix, the uterus, and the vaginal and perineal areas can result in severe pain. However, at the end of this stage, the baby is born! The third state is the expulsion of the placental membranes, which is usually completed within 30 minutes.

Twins are born about once in every 80 births. Nonidentical twins result from the fertilization of two different eggs by two different sperm, at roughly the same time. These twins will have different genetic make-ups. Identical twins result from a single egg fertilized by a single sperm, with the fertilized egg then dividing into two individuals. These twins will have similar genetic characteristics.

Now let's explore the formation of life and the microscopic world of cells.

The Cells, DNA, and Tissues

An Electron Microscopic View of a Cell

Mitochondria are cell organs located within the cytoplasm and are involved in functions that manufacture products that provide energy needs for the cell.

The cell membrane is the outer covering and protects and allows certain materials to enter the cell.

The cytoplasm consists of all the other material within the cell membrane, but bounded by the nucleus.

The Golgi apparatus also lie within the cytoplasm and are important in producing and storing secretions.

The nucleus carries the genetic material of the cell and allows it to reproduce.

Energy is stored within the cytoplasm in the form of glycogen and lipids, which are utilized as needed.

DNA and Chromosomes

DNA (deoxyribonucleic acid) is the basis for the genetic makeup of your body. It controls the activities of the cells and the functioning of the whole body.

DNA consists of two parallel chains traveling in opposite directions as a double helix. Each chain joins together multiple sugar (S) and phosphate (P) components. The two chains are joined together at their sugar molecules by bases (B). RNA (ribonucleic acid), also a linear chain, translates information from the DNA into other proteins, and thereby provides information to other cells and other parts of the body.

Chromosomes are rod-shaped bodies containing genes of a linear sequence of DNA and other proteins. Each cell contains 46 chromosomes, of which there are 23 pairs—one pair obtained from each parent. Medical disorders of chromosomes can result from several factors: a decrease or increase in the number of chromosomes or an abnormal arrangement of the chromosomes; an abnormality in the information in one or more chromosomes; or the interaction of a chromosome's genes and the environment. For example, there may be a genetic tendency for high cholesterol; if this tendency is coupled with a high cholesterol diet, the cholesterol level may be higher than what it would be if the genetic tendency were not present.

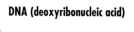

DNA (deoxyribonucleic acid)

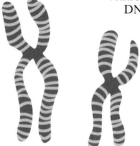

Human chromosomes

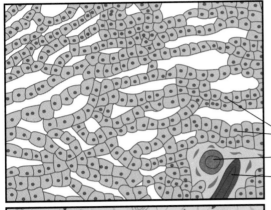

The Body's Four Types of Tissues

Epithelial tissue   Epithelial tissue consists of layers of cells covering the outer body surfaces (like the skin), the inner body surfaces (like the inner cells of the lungs, stomach, or bladder), and the cells of glands (including the liver, pancreas, and ovaries). Some epithelial tissue, like the liver tissue shown here, has cells that can vary in shape and may have more than one nuclei.

- Liver cells
- Artery
- Vein

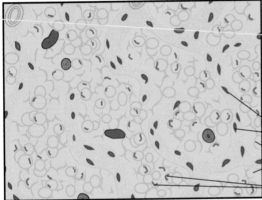

Connective tissue   Connective tissue is widely found throughout the body and is usually beneath the skin. Examples of connective tissues include fat, cartilage, bone (such as seen here), blood, and lymph. This figure depicts bone as it would be found in the sternum (breastbone).

- Bone cells
- Marrow cavities with marrow and fat
- Fat cells

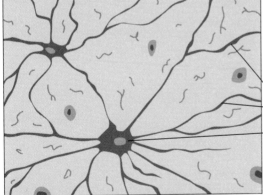

Nerve tissue   This figure shows a particular type of nerve tissue that is highlighted by the presence of astrocyte cells. Astrocytes are star-shaped cells with leglike processes that have a variety of roles in the nervous system, namely transfer of nutrients and healing and scar formation.

- Processes of astrocytes
- Cell body and nucleus of astrocyte

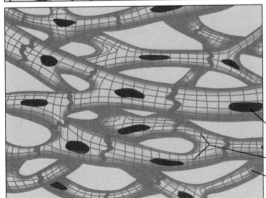

Muscle tissue   In general, muscle fibers are elongated cells with shapes that help them shorten or contract. There are three basic types of muscle tissue: skeletal or voluntary (as in the muscles of your arms or legs), smooth or involuntary (as found in your stomach or uterus), and heart muscle. Heart muscle fibers, which are shown here, are usually larger than smooth muscle and have cross-striations, branching, and cells with large oval nuclei in the center of the muscle.

- Nucleus of heart fibers
- Branching of heart fibers
- Cross-striations

The Formation of a New Life

1 A new life begins with the fertilization of an egg by a sperm in the fallopian tube.

2 The fertilized egg then begins dividing into progressively larger cell masses—starting with 2 cells, then 4 cells, then 8 cells, until it forms a cluster of cells called a blastocyst.

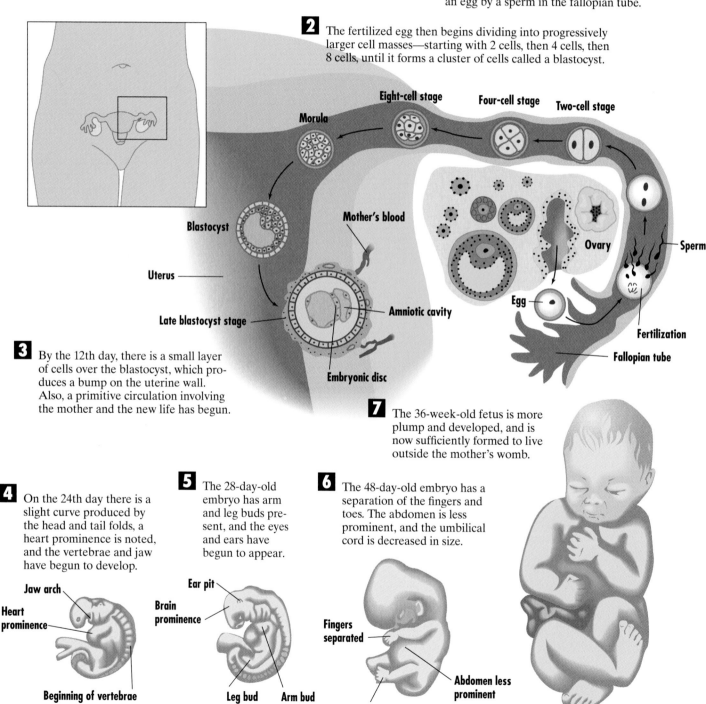

Eight-cell stage

Four-cell stage

Two-cell stage

Morula

Mother's blood

Ovary

Sperm

Blastocyst

Egg

Uterus

Fertilization

Amniotic cavity

Fallopian tube

Late blastocyst stage

Embryonic disc

3 By the 12th day, there is a small layer of cells over the blastocyst, which produces a bump on the uterine wall. Also, a primitive circulation involving the mother and the new life has begun.

7 The 36-week-old fetus is more plump and developed, and is now sufficiently formed to live outside the mother's womb.

4 On the 24th day there is a slight curve produced by the head and tail folds, a heart prominence is noted, and the vertebrae and jaw have begun to develop.

5 The 28-day-old embryo has arm and leg buds present, and the eyes and ears have begun to appear.

6 The 48-day-old embryo has a separation of the fingers and toes. The abdomen is less prominent, and the umbilical cord is decreased in size.

Jaw arch

Heart prominence

Beginning of vertebrae

Ear pit

Brain prominence

Leg bud

Arm bud

Fingers separated

Toes separated

Abdomen less prominent

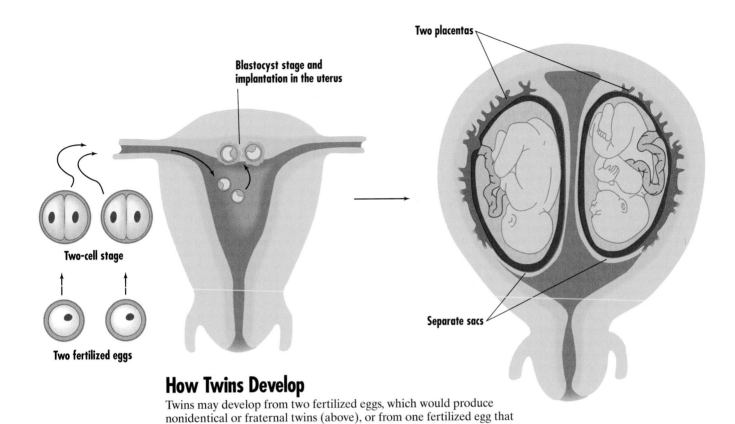

How Twins Develop

Twins may develop from two fertilized eggs, which would produce nonidentical or fraternal twins (above), or from one fertilized egg that would produce identical twins (below). Twins occur in about 1 out of every 80 or 90 pregnancies, and about two-thirds are identical twins.

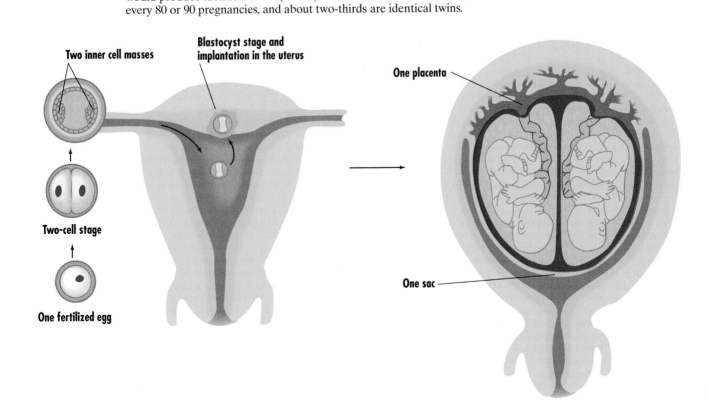

HOW your Body works

QUICK START CARD

System Requirements For MPC CD-ROM

How Your Body Works requires the following system components:
PC 486 SX25 (486 SX33 recommended)
8 MB RAM
Windows 3.1
Double speed CD-ROM drive (300 KB/sec transfer rate)
Super VGA 640x480
256 (8 Bit) color [65,000 (16 Bit) color recommended]
4 MB available on hard drive (10 MB available recommended)
Sound Blaster or compatible sound card
Mouse

* Low Memory Detection on Boot

FAMILY FEATURE: VISUALLY SENSITIVE MATERIAL ON HUMAN SEXUALITY CAN BE MADE INACCESSIBLE AT INSTALLATION.

Installation

The program automatically installs Video for Windows and *How Your Body Works* on your system as shown below.

Note: If you already have Video for Windows, in step 4 below you can type D:\HYBW\SETUP.EXE and press Enter to set up How Your Body Works.

1. Put the disc in your CD-ROM drive with the label side up.
2. Run Windows.
3. Open the Program Manager **File** menu and click **Run**.
4. In the Command Line, type **D:\INSTALL.EXE** and press **Enter**.
Note: If your CD drive is another letter, substitute it for D. For example, type **E:\INSTALL.EXE** and press **Enter**.
5. At the **How Your Body Works Installer** dialog, click **OK**.
The program installs Video for Windows, restarts Windows, and opens the **Installation Options** dialog box.
6. Select your install options and click **Continue**.
Note: Installation defaults so the Human Sexuality files are turned on. To block the installation of Human Sexuality topics, check the box.
7. Follow the onscreen prompts to complete the installation.

Launching the Program

To launch *How Your Body Works*:
1. Put the disc in your CD-ROM drive.
2. Run Windows and double-click the program icon
3. Click each legal screen to begin. The door appears.
4. Click the door to enter the **Lab**.
Note: To find about the program (and the Scavenger Hunts), click the **Welcome** note pinned to the door.

The Lab

The **Lab** is your main menu. When you first enter the Lab, the telephone is ringing. Click on the phone to get help or click another feature to use it as shown below. Move your mouse around to find hotspots (pop-up labels appear or your cursor changes). Click a hotspot to use that feature.

How Your Body Works includes the following features. Click the indicated area on the Lab to open each feature.

Feature	Where to click
Help	*Telephone*
Anatomy	*X-rays*
Anatomy Chart	*Wall Chart (to left of X-rays)*
Body Tours	*Body Model in case*
Browser (includes print function)	*Computer on Desk*
Disorders	*Desk File Drawers*
Medicine & First Aid	*Medicine Cabinet*
Nurse's Notebook	*Radio on Bookshelf (shelf 3)*
Reference Books	*Books on Bookshelf (shelf 2)*
Related Concepts (includes print function)	*Books on Bookshelf (shelf 4)*
Scavenger Hunt Game	*Notes (there are 3)*
Wellness	*TV/VCR*
Ziff-Davis Books	*Books on Bookshelf (shelf 3)*
Fun Items	*scattered about*

Hint: Look for fun surprises disguised around the room.

To return to the **Lab** from a feature, click the gray area outside of the feature's art.

To return to the desktop, click the **Exit** sign. The credits will be followed by the FYI disclaimer screen. Click to exit.

Help

Click the desk **Telephone** to open help.

To find out how to use a feature, click a topic shown on the telephone.

To return to the **Lab**, click gray areas anywhere outside the help art.

Anatomy

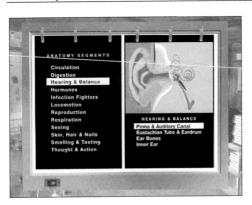

Click the **X-rays** on the Lab wall to study Anatomy.

Click a topic on the left side of the screen to study it on the right side.

Click on a topic listed on the right side of the screen to play the anatomy segment.

To return to the **Lab**, click gray areas anywhere outside the Anatomy art.

Anatomy Chart

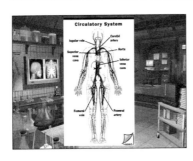

Click the **Chart** on the wall to the left of the X-rays in the Lab to study the Anatomy Chart.

To turn the pages of the chart, click the curled bottom edge of the page.

To return to the **Lab**, click gray areas anywhere outside the Anatomy Chart art.

Disorders

Click the desk drawer to open the **Disorders** files.

Click a **File Tab** to open the information about disorders in that category. Descriptions of disorders in that category appear. Click an index card to play the video segment.

To return to the **Lab**, click gray areas anywhere outside the Disorders art.

Nurse's Notebook

Click the radio on the third shelf of the bookshelf to open the **Nurse's Notebook** radio program.

Click the + or - sign on either side of the knob to change the topic. The title appears on the radio. Click the square button to forward to the next topic which will automatically play. Click the knob to stop and rewind audio.

To return to the **Lab**, click gray areas anywhere outside the Nurse's Notebook art.

Reference Books

Click the books on the middle shelf of the bookshelf to open the References. Two books appear onscreen, a **Medical Dictionary** and **Health Directory**.

Click a book to open it. Click an alphabetical tab to jump to that part of the book. Click the curled edge at the bottom of each page to turn it.

To return to the **Lab**, click gray areas anywhere outside the References art.

Scavenger Hunt Game

The pathologists in the Lab have invented three special devices that reveal the human body in unique ways. But they are secretive scientists. To find these hidden devices, you'll need to go on scavenger hunts to find all the necessary clues. There are three notes: 1) near the microscope, 2) on the high shelf, and 3) in the trash can under the desk. Before you enter, check out the note on the door to the Lab for details!

Wellness

To view the videos, click the **TV/VCR** monitor.

Click a video (stacked to the right of the screen) to view its table of contents on the TV/VCR screen. Click a screen title to listen to a medical expert discuss that topic. To find out more about the experts, click on the book under the stack of videos.

To return to the **Lab**, click gray areas anywhere outside the TV/VCR art.

Browser

Click the **Computer** (on the desk) to use the **Browser**. From the Browser you can read, view, or print text or narrative from topics in the program. The Browser also has its own **Help** button.

Click **System** or **Section** to list the twelve body categories or the special features, respectively. Double click a listed topic to view selections. Click a listed selection to view a video or read text. If you see a video, click **Read** to view the related text.

When text is displayed on the right side of the Browser screen, you can click the **Print** button to get hard copy of that information. And, you can print narrative from any of the animation/video segments.

To return to the **Lab**, click gray areas anywhere outside the Browser art.

Body Tours

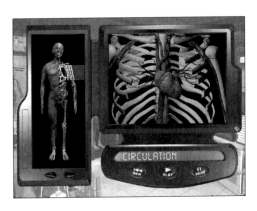

Click the body in the case to view **Body Tours**. On the left side of the screen, click a body part to load that Body Tour. On the right, click to use the command buttons:
- **Rewind**.
- **Play** starts or resumes the tour.
- **Pause** stops the tour.

To return to the **Lab**, click gray areas anywhere outside the Body Tour art.

Medicine Cabinet

Click the **Medicine Cabinet** to open it.

Click a shelf to view information. The top shelf is about prescription drugs. The second shelf contains information on over the counter drugs. The third shelf has medical instruments. The bottom shelf lets you use the **First Aid Manual**. There are also hidden fun surprises on the shelves.

To return to the **Lab**, click gray areas anywhere outside the Medicine Cabinet art.

Technical Support

Before calling for technical support, please have the following information ready:

- Detailed description of your problem and the exact steps described in order of occurrence, so we can try to reproduce the problem.
- Name/make/model/speed of your computer.
- Copy of your CONFIG.SYS file.
- Copy of your AUTOEXEC.BAT file.
- Name/make/model of your sound card.
- Name/make/model of your video card.
- Print out of a memory dump (to do this type MEM/C >[FILENAME] and then print or E mail this file. [FILENAME] can also include a path such as A:\MEMORY.TXT).
- Name/make/model of printer you are using, and whether it is color or black & white.
- Version of Windows and MS-DOS you are using.
- For Windows and MPC products, we also need a copy of your WINDOWS.INI and SYSTEM.INI files, both of which are in your Windows subdirectory.

VERY IMPORTANT: WHEN MAILING IN YOUR PROBLEMS, SUGGESTIONS, OR QUESTIONS, PLEASE INCLUDE THE FOLLOWING INFORMATION:

- Fax number.
- Work phone number.
- Home phone number (or where your computer is most of the time).

For technical support in the USA, please contact:

Mindscape, Inc.
60 Leveroni Court
Novato, CA 94949
FAX: (415) 883-0367
Telephone: (415) 883-5157
BBS   (415) 883-7145
Automated 800 Service (800) 409-1497. This service can help you to find your own answers at no charge!
America Online keyword: MINDSCAPE
CompuServe: GO MINDSCAPE

For technical support in Europe, please contact:
Technical Services
Mindscape International Ltd.
Priority House, Charles Avenue,
Maltings Park, Burgess Hill,
West Sussex, RH15 9PQ
England, United Kingdom
When calling from outside the UK:
FAX: <International Code> 44 1444 248996
Telephone: <International Code> 44 1444 239600
When calling from inside the UK:
FAX: 01444 248996
Telephone: 01444 239600
(Monday - Friday, 09:30 - 13:00 hours and 14:00 - 16:30 hours)

For technical support in Australia and New Zealand, please contact:
Mindscape, Inc.
5/6 Gladstone Road
Castle Hill, New South Wales
Australia 2154
FAX: 02 8992348
Telephone: 02 8992277

Ziff-Davis Press Survey of Readers

Please help us in our effort to produce the best books on personal computing.
For your assistance, we would be pleased to send you a FREE catalog
featuring the complete line of Ziff-Davis Press books.

1. How did you first learn about this book?

Recommended by a friend . . . . ☐ -1 (5)

Recommended by store personnel . . . . ☐ -2

Saw in Ziff-Davis Press catalog . . . . ☐ -3

Received advertisement in the mail . . . . ☐ -4

Saw the book on bookshelf at store . . . . ☐ -5

Read book review in: _____ ☐ -6

Saw an advertisement in: _____ ☐ -7

Other (Please specify): _____ ☐ -8

2. Which THREE of the following factors most influenced your decision to purchase this book? (Please check up to THREE.)

Front or back cover information on book . . . ☐ -1 (6)

Logo of magazine affiliated with book . . . . . . ☐ -2

Special approach to the content . . . . ☐ -3

Completeness of content . . . . ☐ -4

Author's reputation. ☐ -5

Publisher's reputation . . . . ☐ -6

Book cover design or layout . . . . ☐ -7

Index or table of contents of book . . . . ☐ -8

Price of book . . . . ☐ -9

Special effects, graphics, illustrations . . . . . . ☐ -0

Other (Please specify): _____ ☐ -x

3. How many computer books have you purchased in the last six months? _____ (7-10)

4. On a scale of 1 to 5, where 5 is excellent, 4 is above average, 3 is average, 2 is below average, and 1 is poor, please rate each of the following aspects of this book below. (Please circle your answer.)

Depth/completeness of coverage   5 4 3 2 1   (11)

Organization of material   5 4 3 2 1   (12)

Ease of finding topic   5 4 3 2 1   (13)

Special features/time saving tips   5 4 3 2 1   (14)

Appropriate level of writing   5 4 3 2 1   (15)

Usefulness of table of contents   5 4 3 2 1   (16)

Usefulness of index   5 4 3 2 1   (17)

Usefulness of accompanying disk   5 4 3 2 1   (18)

Usefulness of illustrations/graphics   5 4 3 2 1   (19)

Cover design and attractiveness   5 4 3 2 1   (20)

Overall design and layout of book   5 4 3 2 1   (21)

Overall satisfaction with book   5 4 3 2 1   (22)

5. Which of the following computer publications do you read regularly; that is, 3 out of 4 issues?

Byte . . . . ☐ -1 (23)

Computer Shopper . . . . ☐ -2

Corporate Computing . . . . ☐ -3

Dr. Dobb's Journal . . . . ☐ -4

LAN Magazine . . . . ☐ -5

MacWEEK . . . . ☐ -6

MacUser . . . . ☐ -7

PC Computing . . . . ☐ -8

PC Magazine . . . . ☐ -9

PC WEEK . . . . ☐ -0

Windows Sources . . . . ☐ -x

Other (Please specify): _____ ☐ -y

Please turn page.

Cut Here

Cut Here

PLEASE TAPE HERE ONLY—DO NOT STAPLE

6. What is your level of experience with personal computers? With the subject of this book?

	With PCs	With subject of book
Beginner	☐ -1 (24)	☐ -1 (25)
Intermediate	☐ -2	☐ -2
Advanced	☐ -3	☐ -3

7. Which of the following best describes your job title?

Officer (CEO/President/VP/owner) ☐ -1 (26)
Director/head ☐ -2
Manager/supervisor ☐ -3
Administration/staff ☐ -4
Teacher/educator/trainer ☐ -5
Lawyer/doctor/medical professional ☐ -6
Engineer/technician ☐ -7
Consultant ☐ -8
Not employed/student/retired ☐ -9
Other (Please specify): _____ ☐ -0

8. What is your age?

Under 20 ☐ -1 (27)
21-29 ☐ -2
30-39 ☐ -3
40-49 ☐ -4
50-59 ☐ -5
60 or over ☐ -6

9. Are you:

Male ☐ -1 (28)
Female ☐ -2

Thank you for your assistance with this important information! Please write your address below to receive our free catalog.

Name: _____

Address: _____

City/State/Zip: _____

Fold here to mail.   2311-14-15

_____

_____

_____

BUSINESS REPLY MAIL
FIRST CLASS MAIL   PERMIT NO. 1612   OAKLAND, CA

POSTAGE WILL BE PAID BY ADDRESSEE

Ziff-Davis Press
ZIFF-DAVIS
ZD
PRESS
5903 Christie Avenue
Emeryville, CA 94608-1925
Attn: Marketing